Social Darwinism and
English Thought

Social Darwinism and English Thought

The Interaction between Biological and Social Theory

GRETA JONES
Senior Lecturer in Politics, Philosophy and History,
Ulster Polytechnic

THE HARVESTER PRESS · SUSSEX

HUMANITIES PRESS · NEW JERSEY

First published in Great Britain by
THE HARVESTER PRESS LIMITED
Publishers: John Spiers and Margaret A. Boden
16 Ship Street, Brighton, Sussex
and in the USA by
HUMANITIES PRESS INC.,
Atlantic Highlands, New Jersey 07716

British Library Cataloguing in Publication Data

Jones, Greta
    Social Darwinism and English thought. –
    (Harvester studies in philosophy).
    1. Sociology – Great Britain – History
    2. Social Darwinism
    I. Title
    301'.0941   HM22.G7

ISBN 0–85527–811–0

Humanities Press Inc.
ISBN 0–391–01799–3

Typeset in VIP Bembo by
Inforum Ltd., Portsmouth
Printed in Great Britain by
Redwood Burn Ltd., Trowbridge and Esher

Contents

ACKNOWLEDGEMENTS

This book owes a great deal to the practical help given to me by the staffs of the libraries of the British Museum, University of London, and London School of Economics. It also benefited, in its initial stages, from the support of Dr Anne Bohm, secretary to the Graduate School of the LSE. I owe a debt of gratitude to Professor Ralph Miliband who started me on research. Professor D.G. MacRae aided me considerably in the course of my work and his wide knowledge of nineteenth-century social theory proved invaluable. The project was discussed at various stages with Dr Paul Bew, Dr Robert M. Young and Dr Margaret Boden. I am indebted to them for their advice and suggestions. I would also like to thank Professor A.J.A. Morris and Professor Robert Gavin for their assistance.

Greta Jones

PREFACE

ONE prominent contemporary philosopher has remarked that much of philosophical discussion is a fight over a word or a phrase — 'the stake in a decisive but undecided battle'. W.J.M. Mackenzie in a recent book[1] has identified social Darwinism as such a term.

The social theorists of the late nineteenth and early twentieth century were as anxious to claim the epithet 'Darwinian' as some modern social theorists are to disclaim it. For example, E.O. Wilson the sociobiologist, in his recent book *On Human Nature* (1978) attempts to distinguish his own theories from the social Darwinist tradition. To many social theorists of the nineteenth century this would have been a surprising outcome.

Nonetheless there has, in recent years, been a revival of interest in biology and society. This has also been accompanied by the increasing use of the term 'social Darwinist' in polemic. Yet it has not been accompanied by any clarification of it. This book is, essentially, a reinvestigation and reassessment of the influence of social Darwinism and also an attempt to analyse the reasons why it has provoked such controversy. It examines not only the historical roots of social Darwinism but also its apparent re-emergence today as 'sociobiology'.

There are indications that social Darwinism has had and still exercises considerable influence. Social Darwinism flowed into many channels, into liberalism, socialist as well as conservative thought, into ideas about race and class, into the debate over the state and social reform. When social thought transformed itself into an academic subject — sociology — the syllabi and question papers undoubtedly reflect the importance of Darwinism. This, of course, is partly the result of the specific interests of the creators of sociology as an academic discipline — of Westermarck, Hobhouse, Wallas, Rivers, Haddon and Seligmann. But nonetheless its influence over

vii

academic sociology supports the view that the conscious sys-
temisation of the discipline leaned heavily upon biology in
general and Darwin in particular.

In addition this study refers to a wide number of texts and
theorists. The aim has been not only to illustrate the ubiquity
of social Darwinism and chart an outline of its dimensions. It is
also to concentrate on certain key ideas rather than on individual thinkers and to illustrate the dimensions of these ideas
by reference to the variety of ways in which they were put to
work.

A conscious effort has been made to avoid the trap of
writing history backwards; in particular to avoid writing a
genealogy of social theories based on contemporary assess-
ments of 'significant' thinkers. This tends to do two things: to
repress those elements of intellectual history which give us
nowadays a bad conscience; and to transform historical figures
into predecessors or to suppress them altogether. Thus it tends
to be forgotten how close much of Galton's work was, not to
modern theories of heredity but to Gall, Lavater and the
tradition of reading character in physical signs. Also forgotten
is how much Darwinism, nowadays seen as representing
introspective, innate instinct psychology, was close to a
psychology of sensation and environmental determination.

This book also moves between social theory at its most
esoteric, political debate and popular culture. This reflects the
indivisibility of much of these elements in intellectual life.
What Herbert Spencer said about evolution and the state per-
colated to a certain level of popular culture. In a similar fashion, nowadays, much of E.O. Wilson's sociobiology reaches
an audience much wider than the learned academic specialist to
whom he originally addressed his work. Earlier social Darwinists quite clearly aimed for this. Both Geddes and Galton
felt sociology should be put to practical social use and both
consciously aimed at a wider audience. Hobhouse's investigations of the origin of morality and rationality were, ultimately,
interventions within a political debate. Neither Wallace nor
Darwin regarded the impact of their theories on a wider
culture as irrelevant. Therefore, although this book is not a
purely political or institutional record of the influence of social
Darwinism it does touch on these areas. It is, in addition,

directed at a number of current political and social debates.

This is necessarily an interdisciplinary study which charts the relationship between politics, biology, philosophy and the social sciences. Social Darwinism was largely the product of the intertwining of these factors and this kind of analysis poses problems. Social Darwinism emerges as a unified, discrete area of thought only if we look at it from one angle — from politics or from the history of the social sciences for example. But to look at it from the point of view of all these disciplines demands not just a consciousness of the final synthesis achieved but also of the contradictions, tensions and different and unique developments in all these areas of thought. It will, therefore, inevitably and rightly emphasise the fragmented, contradictory character of social Darwinism — the very real hiatus which existed at many points between the progress of biology and the demands of social and political thought.

In spite of this complexity of social Darwinism, there are identifiable organising elements. One of these is politics. Those who have recently attempted to revise Richard Hofstadter's discussion of the social role of biological ideas in defending American capitalism in the late nineteenth century, have missed the point.[2] It is perfectly possible to show, as they do, (and as Hofstadter himself did) a form of social Darwinist thought hostile to competitive capitalism and preaching a milder and more pacific form of society. This book will also show the great variety of political interpretations to which social Darwinism lent itself in the British context. Nontheless, in politics some ideas survive and grow and others do not. There were political and social limits on the 'free market' of ideas. Cultural and social forces picked out some among the litany of social Darwinism and neglected others. So that it is possible to identify, even given the wide spectrum of social Darwinist ideas represented in British and American society, a certain political direction which social Darwinism took as far as its general social influence is concerned.

The second organising element is philosophy. Running through social Darwinism were certain persistent ideas about the philosophy of the social sciences and the philosophy of biology. The two are interconnected. Philosophy originally encouraged the social sciences to look to biology or nature, in general, for their principles (see Chapter I). It also set out the limits and presuppositions upon which social theory could appropriate biology. In the process it frequently intervened within biology itself to make its character more amenable to social theory. It insisted upon a number of ideas. First, that all biological ideas about human society are reducible to a theory of the character of man, and social life to an expression of individual activity. The most appropriate biology was one, therefore, which explained the origin of human faculties. These faculties were delineated by certain assumptions about their role in social behaviour and culture. The evolutionary history of man was to show how those qualities considered most essential to social life emerged. Second, because it saw society as a product of the exercise or failure to exercise these faculties, it demanded to see them at work also in biology. Many of the attempts to revise the biology of man by injecting into it an element of choice— particularly over the direction of human heredity — came from social thought. The disagreements between social theorists and between biologists over the degree to which biology was deterministic arose, largely, because of a conception they had of the basis of social life. To relate biology to social theory implied a recasting of biology to fit the notion that faculty and its exercise was at the bottom of social structure. This tension between the precepts of social theory and the actual structure of biology is at the heart of the history of social Darwinism from the mid-nineteenth century to the present day.

From the point of view of natural science Darwinism is primarily the theory of natural selection. Briefly, this consists of the propositions that, firstly, individuals tend to inherit characteristics from their parents, but also that they vary from time to time. These variations may be inheritable although some (phenotypic variations) are not. Both types, hereditary variation and phenotypic, may enhance an individual's chance of survival. The theory of natural selection adds a number of other propositions to this. Each individual lives in an environment composed of individuals of the same species, other species, a geographical and climatic area, a particular set of food resources etc. — in other words an ecology. All of these may exert pressure upon its survival. The most famous pressure associated with Darwinism is that of population increase, but there are others. Under the influence of these factors, those variations which give an individual a better chance of surviving and leaving more descendants will tend to be perpetuated (selected is the word most often used). This process has led to certain evolutionary changes and, in particular, has enabled new species and varieties within species to emerge.

This is by no means all that can be said about natural selection. Since the emergence of the Mendelian theory of heredity in the early part of this century, the theory of natural selection has incorporated the principle of genetics. A number of scientists in this period showed how elements of Mendelism complement Darwin's original theory and, in fact, solve a number of the problems which, originally, seemed to Darwin's critics to be formidable obstacles to the application of the theory. Natural selection has also influenced other areas in the biological sciences — for example, ecology — and the study of animal behaviour.

There are other elements in Darwin's work. He wrote on geology, made studies of the morphology, physiology and behaviour of a number of species, wrote on psychology and human origins and to a lesser extent, on social evolution. He attempted, unsuccessfully, to develop a theory of heredity. Some though by no means all of these theories have proved valuable in natural science although their value cannot be compared with Darwin's great achievement — the theory of natural selection. Some are now merely historical curiosities.

To Darwin's name must be added that of his contemporary Alfred Russel Wallace who developed a similar theory in the 1850s. Although Darwin's discovery of natural selection preceded Wallace's by probably twenty years, it was Wallace's communication of his discovery to the Linnean Society in 1858 which forced Darwin finally to publish his own work. He did this jointly with Wallace in papers read before the Linnean Society in 1858 and, in 1859, in greater length in the *Origin of Species*. Although Darwin and Wallace differ in certain emphases in the theory of natural selection, Wallace's views basically outline the same premises. Whilst Wallace himself gave the major credit to Darwin in creating the theory, and this book will talk primarily of Darwin and Darwinism,

Wallace's contribution will not be forgotten.

Darwin's view of philosophy was rather negative. This is encapsulated at its most extreme in his opinion that 'He who understands baboons would do more toward metaphysics than Locke'.[3] But when he and Wallace tried to solve some problems, for example, those of human mental development, they found themselves borrowing from philosophy. Darwin, whilst opposing quite unequivocally one set of philosophical presuppositions — about human uniqueness, the innateness of ideas and the divine creation of man — borrowed from the British empirical tradition the notion of the role of sensation and experience in creating ideas.

Thus, Darwin's desire to escape philosophy was not altogether satisfied. These philosophical ideas, however, were not enshrined in the heart of the theory of natural selection. On the contrary, both Darwin and Wallace remained adamant about the character of the theory itself and were not inclined to elaborate on it with 'improvements' or philosophical reinterpretations. Their philosophical positions arose from their confrontation with the question of human development and social evolution. This illustrates the importance of the philosophy of man and society in fragmenting Darwinism and creating social Darwinism. In other words, what was undoubtedly a unique historical event in the biological sciences, was also accompanied by many other processes; its reinterpretation, largely through its extension to man and society; the intervention of certain philosophical ideas within this aspect of Darwinism; and the fragmentation of Darwinism itself — into natural selection, a theory of human development and a philosophy of social evolution. The particular cultural and historical milieu around Darwin did not ultimately decide the role of the theory of natural selection in biology but it did influence the course of its development and above all the character of its application to man.

This history will assume certain things about scientific development which have implications for the study of history in general. The first casualty of it will be the notion that there is a clear unilinear line of discovery and development in the biological sciences from Darwinism to Mendel, from there to population genetics and eventually to the discovery of DNA.

In one sense this history does exist, but if we want to study the actual historical events surrounding the emergence of social Darwinism, then the state of the biological sciences at that particular time must be taken into consideration. This included the problems and grey areas as well as the achievements and clear advances. To a large extent the history of social Darwinism is the history of how the *problems* — rather than the achievements — in Darwinism were put to work in the interests of social thought. In addition, biological science in the nineteenth and twentieth centuries was composed of two different sets of ideas: those which can clearly be identified from the standpoint of contemporary science as concepts which are still operative in modern science and those which are not. The latter include a wide range: certain philosophical notions, tentative solutions to genuine problems which were later discarded and also social preconceptions. Certain histories of science write these out — but for the kind of history relevant to this study they are an essential component. They often represent the 'shared' areas between the intellectual discipline of science and social thought, and they form the basis upon which much of that relationship was built up.

Finally, another influence helped create biological sociology. A great deal of pre-Darwinian social thought was a homily upon the inevitability of society and nature taking the form they did. Paley's *Natural Theology* (1802) and the *Bridgewater Treatises* (1835) were attempts to reconcile the observations of science with a pre-ordained nature. This notion of order and design in nature was paralleled by similar conceptions about social order. These exegeses were, of course, also defences of the role of God in the natural and social world.

Social Darwinism secularised these ideas. It removed God but it reinstated the idea of order, equilibrium and hierarchy, this time in a social context. It therefore 'naturalised' the social order. Even the social Darwinist's interest in evolution was often basically a journey into the past in order to discover the roots of their own society. Social Darwinism substituted natural, scientific processes for God as the guarantor of social equilibrium. This had profound consequences for the status of science in society. Much of contemporary struggle over the meaning of science for and against its influences arises precisely from its use for this purpose. How Darwinism — the theory of change and disequilibrium in nature — came to play in social theory this paradoxical role is the subject of the first chapter.

I THE MORAL ECONOMY OF NATURE

IN 1859 Charles Darwin published *The Origin of the Species* a work which set out the case for evolution in natural organic life and a mechanism, natural selection, by which this evolution took place. This did not, as has been the traditional view of some historians of social thought, inaugurate a new epoch in social theory. It did, however, give rise to certain important developments: a generation of social theorists who, for example, explicitly claimed to have based their work on the Darwinian theory of natural selection.

Before the *Origin*'s publication there already existed a readiness to accept the relevance of biological thought. Comte claimed that, 'the subordination of social science to biology is so evident that nobody denies it in statement however it may be neglected in practice'.[1] Comte maintained that the biological sciences were the immediate historical precursors of sociology and the logical base upon which the theories of the social sciences could be built. Social Darwinists frequently repeated this idea. Benjamin Kidd claimed that 'History and politics are merely the last chapters of biology— the last and greatest— up to which all that has gone before leads in orderly sequence'.[2] Bagehot believed that 'it is most important that there should be no disunion between political economy and physiology or between it and the more complex forms of social science'.[3]

The idea of a connection stemmed from two basic related ideas: first, that sociology was the science of the human subject and that biology was, at least in part, a science of faculty and behaviour; second, that there were analogies between natural and social life. In regard to the first idea D. G. Ritchie, the philosopher, described it as a case of 'whatever else human beings may or may not be at least they are living beings subject to the laws of biology'.[4] If biology threw light upon human attributes inevitably it was of interest to sociology.

How persuasive this notion was is shown by certain currents of thought in the social sciences. Alfred Marshall, the economist, who contributed the notion of marginal utility to economics, was still thrown back upon what was basically an anthropology when describing economic behaviour. Economic progress was, he wrote in *The Economics of Industry* (1879), a product of character and faculty. Man 'improves' the gifts of nature or 'wastes' them and the basic means he uses or neglects to use were '(1) his physical strength and energy; (2) his knowledge and mental ability; (3) his moral character'.[5] To the extent that economics was a science of psychological attribute, then biology was considered relevant to it.

One of the impulses pushing the social sciences towards biology was this idea of the possibility of a science of human character which would include both a knowledge of man's physiology and social behaviour. This did not necessarily confine biology to the provision of simple descriptions of mental and physical faculty. Comte believed that biology could also provide the key to certain aspects of human moral character. To the distaste of John Stuart Mill, in the *Cours Positiv*, Comte singled out the work of the phrenologist Gall as particular interest to sociologists. Comte thought it possible to identify human character and personality by certain human physiological features. For Comte, the cerebral localisation of faculties was the most truly scientific branch of psychology. This was because it 'unites our positive knowledge of the soul with that of the body'. According to J. S. Mill, this view would make 'Mental Science . . . a mere branch, though the highest and most recondite branch of the Science of Physiology. This is what M. Comte must be understood to mean when he claims the scientific cognizance of moral and intellectual phenomena exclusively for physiologists.'[6] Mill preferred a more introspective approach to psychology. But others in the nineteenth century accepted the view that the investigation of moral and mental character might become part of physiology.

The other connection between the biological and social sciences was the adoption of analogies. The possibility of the use of analogous concepts of biology in sociology was, to some extent, justified on similar premises to those justifying the connection between biology and the science of human behaviour. Both biology and social theory, according to

Comte, were sciences of 'the vital order', that is, life. The similarity in subject matter justified the application of analogous concepts. It was also supported by another and characteristic assumption of nineteenth-century philosophy: that general laws of development applied across a wide range of different sciences. Marshall described this as a process by which 'Physical science has learnt that an increasing knowledge of variety and complexity of the phenomena of nature has often been accompanied by a dimunition of the laws required to explain them.'[7] These laws illustrated a fundamental unity of action between the laws of nature in the physical and moral world.

It was the laws of physics rather than biology that provided the exemplification of scientific method to which other sciences ought to aspire. T. H. Huxley, regarded biology as, ultimately, an amalgam of physics and chemistry. Similarly Herbert Spencer's laws of social and organic evolution were intended to express a physical principle — the persistence of force. This conviction of the possibility of cross-disciplinary general laws and of the close connection between biology and sociology supported the transfer of concepts and ideas between them. At certain points, it justified an even wider dissemination of biological concepts throughout the other sciences. Merz, the historian of nineteenth-century thought, regarded, 'the introduction of the general formulae of selection, adaption and evolution as only another instance of the tendency . . . to look upon the actual things and phenomena of nature merely as examples of general processes':[8] To support this, he cited examples of the use of the idea of natural selection in physics. In his opinion, as in Marshall's, science was, in any case, tending towards general and all-embracing laws.

There was no prejudice against the application of pre-existing hypotheses to the investigation of society. Social investigation in the nineteenth century displayed an exaggerated respect for the collection of pure data. But also characteristic of it was the highly theoretical corpus of work found in Spencer and Comte. This division, however, was not necessarily regarded as antagonistic. J. S. Mill argued that sociology was not a deductive science. There could be no experimentation in the social sciences due to the character of its subject matter. Comte argued that empirical investigation must take place within an existing framework of laws. Mill called these the 'middle principles' of sociology — linking the facts of social life and the highly theoretical character of the more developed sciences from which sociology drew guidance.

Mill, Comte and Spencer were all convinced of the need for a framework for social investigation. They believed that sociology was a body of knowledge in search of the 'middle principles' Mill described. Comte and Spencer attempted to provide these in part by drawing upon biological science for a science of character and behaviour and for analogies between social processes and living organisms. There would, therefore, have been a conscious search for a biological underpinning to the social sciences, even if Darwin had never existed. Social Darwinism was to a great extent a product of existing assumptions about the character of sociology.

This does not, however, dispense with the question of why Darwinism assumed special importance. Darwin was rapidly cast in the role of midwife to the social sciences. W. H. Mallock in 1894 described the effect of natural selection in these terms. Darwinism, he claimed:

provided sociologists with a fundamental and general principle by means of which the sequence of social changes could be seen, represented and explained as possessed of some continuous meaning; and, above all, it supplied them, by its account of the struggle for existence, with a theory which enabled them to reduce to some common and intelligible process the apparently endless varieties of social change and action. It seemed as though suddenly it had made social science vertebrate, giving it some framework round which to group its details;[9]

Mallock over-exaggerated Darwin's influence on social thought — an influence of which, incidentally, he disapproved. As J. W. Burrow has pointed out in *Evolution and Society* (1966), evolutionary social theory had already emerged quite independently of Darwin. The reason for the exaggeration of Darwin's influence lay in the expectations which the philosophy of the social science had created. Even if Darwin had not inspired all evolutionary thought, according to the philosophy of social science he ought to have. Here was the general synthesiser who had made biology evolutionary and above all implied that the origins of man himself could be subject to a scientific investigation. Many of the popularisers of Darwin found the temptation irresistible to rewrite the history of nineteenth-century science to conform to these expectations. Grant Allen, for example wrote:

the evolutionary method has invaded each of the concrete sciences in the exact order of their natural place in the hierarchy of knowledge. It had been applied to astronomy by Kant and Laplace before it was applied to geology by Lyell; it had been applied to geology by Lyell before it was applied to biology by Darwin; before it had been applied to biology (in part at least) by Lamarck and the Darwins, before it was applied to psychology by Spencer; *and it is only at the very end of all that it has been applied to sociology* and the allied branches of thought . . . [10]

This kind of interpretation was popular enough to pose problems for these social theorists who, though they were evolutionary, claimed independence from Darwin. Tylor felt obliged, in the foreword to the second edition of *Primitive Society* (1871), to disclaim explicitly the influence upon his work of either Darwin or Spencer. However, this view was sufficiently influential towards the end of the nineteenth and at the beginning of the twentieth century to create a historiography of the social sciences which repeated the rather inflated claims for Darwin's influence. But although the philosophical background to Darwin may have led to a misrecognition of his impact, Darwin's work did have influence. Those who have asserted otherwise are committing the occupational error of the revisionist historian of throwing out the baby with the bathwater.

Like many sciences, the theory of natural selection developed using a language it was forced to share with other areas of knowledge and belief. Darwinism shared a vocabulary with social thought. It was, therefore, 'recognisable' to social theorists in certain specific ways and had links with a much broader culture than other nineteenth-century scientific developments. Whereas in many ways this increased its impact, it also gave rise to difficulties.

Before Darwin, survival of the fittest was a widely held notion about the natural and social world. The Reverend William Kirby in the *Bridgewater Treatises* of 1844 had described both the human and natural world in these terms. 'The King of the Terrestrial globe man', he had written, 'constantly engaged in a struggle with his fellow man, often laying waste the earth . . . his subjects of the animal kingdom following the example of their masters and pitilessly destroying each other — the strong oppressing the weak'.[11] Conventional opinion found this view perfectly compatible with an anti-evolutionary position. According to John Crawfurd of the Ethnological Society of London, 'As to "the struggle for life", there is no doubt but that, through all living beings, it is the weak that perish and the vigorous that survive. Nature, in some cases, takes some pains for *preserving the integrity of the species but never for its improvement by mutation.'*[12] The struggle for existence was considered to be, by many, an argument against species change. It also had social implications as well as an explanatory role in natural history. Malthus's theory of human society used very much the same idea of a natural equilibrium in which change — in his case population pressure — served only to reinforce the existing relationship between the distribution of people and resources.

Herbert Spencer's articles of 1852 in the *Westminster Review* were directed precisely at the 'static' character implied by this view of the struggle for existence. Although Spencer's article has been seen as anticipating Darwin, it perpetuated the notion, already held by Kirby and Crawfurd, that elimination through struggle preserved the 'type' of the species. Spencer argued only that the struggle, by eliminating the impure specimens of a race, led to constantly improving 'type'. This was not transmutation of the species; rather, immanent evolution. Out of the existing material of the race, the 'ideal' of that species was gradually attained. But Spencer had demonstrated, to his own satisfaction at least, that the struggle for existence could lead to evolutionary change.[13]

This was only one solution proposed to circumvent the anti-evolutionary character of much of social and natural philosophy. Many saw the *Origin of Species* largely in terms of the defection of a prominent scientist to the ranks of evolutionary thought. This tended to disguise the distinctiveness of Darwin and Wallace's theory. In contrast to Spencer, Darwin thought evolution was more than the realisation of the archetype. By rooting the process of evolution in organic variations he suggested that the notion of an 'ideal' or 'type' of a species, was, in any case, nonsense. The 'unfit', in the sense of the variation from a supposed archetype of a species, might very well become the successful progenitor of a new one. The species could only be measured against its ability to propagate its kind not against any idealised version of its essence and character.

But if Darwin moved away from this idea, he appeared to share another concept with Spencer. Malthus's theory of the pressure upon subsistence caused by the growth in population was cited by Darwin in his *Autobiography* as the final inspiration which made the formulation of the theory of natural selection possible. However, population pressure was only one of the mechanisms which could produce evolutionary change. For Darwin population pressure was one of the strongest impulses towards evolutionary change, but it was also the result and not merely the cause of the selective process. In contrast to Malthus, population increase promoted evolutionary change rather than inhibited it. Malthus had seen population pressure as a conservative force leading to the preservation of society in its present form. Darwin treated it as a force for change. Very few social Darwinists who used 'extinction' and 'competitive struggle' were able to make a successful transition from the old to the new conception. W. R. Greg, the Manchester economist, a decade after the publication of the *Origin* still talked in terms of 'that righteous and salutory law of natural selection in virtue of which the *best specimens of the race . . . continue the species and propagate an ever improving and perfecting type of humanity*'.[14]

Francis Galton used natural selection in this sense as did Karl Pearson. Although both had a better appreciation of it as a biological theory than many other social Darwinists, they frequently lapsed into talking of it in this way when they broached the problem of human and social evolution. In 1912, Pearson wrote of natural selection as a theory in which 'the individual better fitted to its environment lived longer than its fellows, had more offspring and these, inheriting its better fitness, raised the *type* of the race'.[15] In addition the interpretation of the role of population as conservative in human society was soon re-established. T. H. Huxley, for example, saw the law of increase as the prime obstacle to social regeneration. In terms very reminiscent of classic Malthusianism, he argued that,

So long as unlimited multiplication goes on, no social organisation which has ever been devised, or is likely to be devised, no fiddle-faddling with the distribution of wealth, will deliver society from the tendency to be destroyed by the production within itself, in its intensest form, of that 'struggle for existence' the limitation of which is the object of society.[16]

Social Darwinism therefore reasserted many of the traditional ways in which survival of the fittest and population pressure had been used before Darwin. Struggle for existence tended to be used as a mechanism by which a social and natural hierarchy was preserved by the distribution to places in it of the 'fittest'. The idea of 'fitness' tended to be imbued with conventional notions of the desirable and valuable. Change and evolution became the means by which ultimate order and the realisation of these 'ideal' faculties and types was achieved.

However, there was an important intellectual shift because of the *Origin* which was unwelcome to many proponents of a natural and social equilibrium. At the very least Darwinism appeared to secularise these ideas. It could be argued that natural theology had already minimised the role of divine intervention in nature by suggesting that the laws of nature — once set in motion by God — could be investigated according to different canons of causality. Indeed one could argue that this concession — intended to accommodate science and religion — could be pushed in the direction of excluding divine intervention from nature altogether. Certainly Darwin removed it from creation, and in his Notebooks on human evolution written in the period 1838–40 speculated on the possibility of removing it from descriptions of human evolution.

In addition, to concentrate too exclusively on the reappearance in 'Darwinian' guise of many of the highly conservative notions of the natural and moral economy — which Spencer revived with such effect in the 1880s — would not do justice to the real changes and developments arising from the *Origin*. Initially at least, Darwin had a disruptive effect upon the existing ideologies of social life.

This became more significant when Darwin undertook to explain human evolution in the *Descent of Man* (1871). The challenge to religious interpretations of nature was implicit in the *Origin*: it became explicit on the publication of the *Descent*.

Darwin drew the materials for a theory which excluded the Creator from human development from a liberal intellectual milieu. It was one often conscious of its opposition to conservative orthodoxy in social and natural thought. In the 1880s the philosopher D. G. Ritchie described this as an alliance between Darwinism, liberalism in politics and anti-intuitionism in philosophy. This, to some extent, would justify Nordenskiöld's — the historian of biology — description of Darwinism as 'the social conceptions of contemporary liberalism (applied) to life and nature', which in his opinion, 'is at once realised from the acknowledged part played by Malthus's social doctrine in the working out of Darwin's theory'.[17] When Nordenskiöld wrote this link had come to mean for many the connection between Darwinism and laissez-faire — an outgrowth of Spencer's interventions in the 1880s. But the relationship between Darwinism and social thought is, however, more complex than this. This is particularly true of the origins of Darwin's views on human mental and social evolution.

II DARWIN, WALLACE AND SOCIAL THEORY

THE influence of Darwin on social theory falls into two kinds. There are, firstly, the comments and ideas which he developed himself. Secondly, there is 'social Darwinism', an elaboration by others of what they consider to be Darwinian interpretations of society. The two are often quite distinct. Nonetheless, it is also true that Darwin did inaugurate a tradition of examining human evolution in a particular way. In the *Descent of Man* (1871) he put forward a theory of social and human evolution. Whilst little of it was original and it brought together materials and explanations which were already available, it passed into intellectual life as Darwinism applied to man and, as such, directly influenced a number of thinkers.

Why did Darwin undertake a theory of human social and psychological evolution? Part of the answer lies in his conception of science. He shared with his contemporaries the belief that a theory must explain all contingencies. Obviously the theory of the evolution of living forms had implications for man. Although Darwin relegated his discussion of man to a few sentences at the end of the *Origin*, a great deal of public debate about the theory of natural selection in the 1860s was about its implication for human evolution.

Darwin avoided extended discussion of evolution and man until the publication of *Descent* in 1871. Interestingly, however, the question of human evolution was discussed very thoroughly in his *Notebooks* compiled in the late 1830s and 1840s.[1] Much of this discussion found its way into the *Descent*.

What the *Notebooks* indicate is that before Darwin was forced into public discussion of man and evolution, he was anxious to solve certain problems of human development, primarily, for his own satisfaction. It could be argued that these discussions — which are less agnostic and more bold than many of his later writings on the subject — were important in the development of the theory of natural selection. This was

10

not because they either led to or were careful derivations from its principles. They were not. But they were attacks upon a constellation of philosophical ideas which made it more difficult to argue for naturalistic causes for human and animal evolution.

To attack these ideas, Darwin in the *Notebooks* and *Descent* compiled various evidences for a naturalistic development of man's social and intellectual characteristics. Most of the ideas he used were not his original contribution, except in the case of the theory of sexual selection which became an important element of animal ethology. But a great deal of his theory of human development was constructed out of generally accessible intellectual ideas of his time. These ideas were arguments against the notion that man and society could only be explained by invoking divine intervention; secondly, against the notion that the gap between the human and animal world was too wide to be bridged by scientific explanation of man's animal origin.

The outline of the problem Darwin set himself was this. Evidence for the discontinuity between man and animal was generally posed in terms of man's intellectual, moral and social attainments. For example, according to one critic,

It is very difficult to see how, by the 'struggle for existence' which is the only motive power that Mr Darwin seems to allow us, the higher intellectual powers of ratiocination, abstraction and self consciousness can ever have been called into action. We can conceive how . . . man might have become more crafty than the fox, more constructive than the beaver, more organised in society than the ant or the bee; but how he can have got the impulse . . . to follow out abstract ideas . . . this is indeed hard to understand.[2]

To these qualities Darwin's critics added language, the aesthetic and moral sense and social organisation. The objection to transmutation was put in two ways — the inadequacy of natural selection itself to explain human development and the inaccessible gap which divided man, given all these characteristics, from the primate. In the opinion of his critics to explain the development of these faculties put too much strain on the fabric of transmutationism and this was one of the reasons why many chose this as an area on which to fight a battle against it.

It was possible, of course, to assert that man's characteristics were natural and could be described by science without demanding that the explanation should be drawn from trans-mutationism. In other words nature might be a rational order in which both man and animals shared a place. Each might have a natural though separate genesis. But Darwin's theory required the establishment of man's descent. He was required by the exigencies of this theory to prove continuity between the complex faculties of man and the animal ancestor.

The *Notebooks* on evolution Darwin compiled in the period 1838–40 and the *Descent* shared these similarities. Both were structured around the necessity to provide a theory of continuity between man and animals. Both were dialogues with potential or existing critics whose objections it is possible to reconstruct from Darwin's texts. They were both therefore trapped within certain pre-existing assumptions of the character of human faculty. This is extremely important in the case of both Darwin's and Wallace's reflections on human social and mental evolution. Darwin, in particular, took upon himself the task of explaining a picture of man painted by a specific set of cultural and social ideas. Not only his critics but his friends held a particular view of 'man' and 'society'. Darwin, therefore, sought a naturalistic explanation for the emergence of essentially an ideological version of contemporary human social and mental evolution. There were, however, certain differences between the context in which the *Notebooks* were written and the *Descent*. The critics who are anticipated in the *Notebooks* materialised in the disputes of the 1860s. The other historical fact intervening is the independent development in the 1850s and 1860s of a natural scientific tradition in describing the development of human faculty and social organisation upon which Darwin drew extensively. In addition Darwin incorporated in the *Descent* elements of the anthropology of the 1860s.

One of the props upon which Darwin's treatment of human social and mental development was built was anthropomorphism. Darwin argued that the germs of complex faculties were found, in an admittedly undeveloped form, in animals. The *Notebooks* in fact were interspersed with anecdotes which set out to demonstrate this. Lord Brougham, for example, was cited as writing, 'on subject of science connected with Nat. Theology . . . says animals have abstraction because they understand signs— very profound— concludes that difference of intellect between animals & man only in kind. probably very important work'.[3] The object of statements like these was set out very clearly by G. J. Romanes whose animal psychology of the 1880s developed this legacy of Darwinism. It was, according to Romanes, to assert that 'The emotional life of animals is so strikingly similar to the emotional life of man — and especially of young children — that I think the similarity ought fairly to be taken as direct evidence of a genetic continuity between them.'[4]

It was not just evidence of genetic continuity that was provided by anthropomorphic descriptions of animal behaviour. It also provided a sketchy prototype of human behaviour rooted in an historical beginning. It then became possible to assert that somewhere in the process of evolution the seeds of human faculty acquired complexity and sophistication. They became in fact the recognisable human characteristics whose existence his critics saw as fatal to the theory of transmutation.

It also meant that the correlations in physical structure between, for example, man and the primates were paralleled by psychological correlations. The history of psychology after Darwin was, in fact, surprisingly similar to that of comparative morphology. Psychologists like Romanes looked for similarities in mental structure between man and animals. In a similar fashion to embryology, psychology treated the mental development of a human child as a recapitulation of the general stages of the race's psychological evolution. They looked for the early psychological forms which had survived frequently finding these in the mental character of the 'savage' races. In the survival of 'instinct' in contemporary man, they felt they had discovered a redundant historical structure alongside more sophisticated mental apparatus. Finally many saw mental collapse as a form of psychological atavism — a throw-back to the primitive mental state, just as many natural historians believed that the domesticated animal would reproduce characteristics of his more primitive ancestors when set free in the wild.

Secondly, Darwin used associationist psychology as the mechanism by which these initial attributes in the animal became the sophisticated apparatus of human faculty. Associationism was of predominant importance in Darwin's psychology. Briefly, associationism explained the origin of ideas as an interaction between the senses and the environment. It presupposed that ideas were, in some way, derived from experience of an external reality. Eighteenth-century philosophy discussed problems about the validity of knowledge obtained in this way and associationism proved an important element in various discourses on moral philosophy and psychology in Darwin's day. In the *Notebooks* he referred often to writers who had been influenced by it.

Darwin was not primarily concerned with the philosophical questions raised by associationism except in so far as it justified a naturalistic explanation of human development. For example, it posited two elements — the human senses and an external world — both of which could be subject to scientific investigation. It therefore suggested that nothing was either preordained or generated by processes which transcended physical reality, man's understanding or scientific examination. However, Darwin was interested, above all, in the psychological aspect of associationism. In the *Expression of the Emotions in Man and Animals* in 1872 for example, Darwin runs through the gamut of associationism, contiguity of ideas and the physiology of the nerve reflex. These psychological conceptions had the overwhelming advantage for him that they were derived from 'natural' processes. They treated man as the product of 'natural' forces, of, for example, the relationship between his physiology and the objective environment. They therefore avoided deriving mental states from intuitive ideas. They dispensed with any mysterious psychical origin for mental character.

The relationships between Darwin and associationism were complex. In the *Notebooks* he used two aspects of it: its materialistic or naturalistic character. Secondly, he applied the notion of contiguity of ideas as the means by which complexity in mental faculty was built up. For example in the *Notebooks* he described the moral sense as arising,

From our enlarged capacity . . . (acting) on strong instinctive sexual, parental & social instincts giving rise to 'do unto others as yourself', 'love thy
neighbour as thyself'. Analyse this out, bearing in mind any new relations
from language . . . . May not idea of God arise from our confused 'causation' in relation to this 'ought'?[5]

The associationism in the *Notebooks* was selected for those
qualities which most strongly opposed it to intuitionism. Of
the other characteristics — for example the examination of
faculties as processes of mind alone (and Alexander Bain, for
example, strongly demarcated associationism from phrenology or other psychologies of physiological determination)
there is little mention. Nor is there much explicit use of
associationism in an evolutionary context. The processes
which Darwin described in the *Notebooks* could be germinated
and developed within the lifetime of an individual. They established that the processes of development of faculty in man and
animals were similar. There was an implicit assumption of
transmutation but no marriage of it with associationism.

This development had to await the independent arrival of
evolutionary associationism. Between 1840 and 1867 — in the
latter year Darwin began to reassemble notes on man for the
writing of the *Descent* — associationism had developed in
several ways which had significance for the project that Darwin returned to in 1867. Briefly these years saw the publica-
tions of Bain (1855, 1859, 1861, 1868), Spencer's *Principles of
Psychology* (1855), Maudsley (1868), and Bagehot (1867) and
(1868). The developments relevant to Darwin were these. First
associationism acquired an increasingly physiological character — not simply the introduction of the notion of nerve reflex
but the idea — taken up by Spencer, Maudsley and Bagehot —
that the reaction of faculty with the environment could produce some memory of itself. It could in fact, be incorporated
into the individual's heredity. In the case of these three writers
this 'memory' was very quickly subsumed into the notion of
use-inheritance. Darwin, although his relationship with use-
inheritance is complex, never completely fell into the habit of
linking it with associationism in this way. He did, however,
begin to assume that the experiences by which faculty was
developed were, in some way, hereditary.

More important than this was the character of associationism which had struck J. S. Mill reviewing Bain's work in 1859. 'Associationist psychology', he wrote in the *Edinburgh Review*,

does in many cases represent the higher mental states as in a certain sense, the outgrowth and offspring of the lower. But in other cases, philosophers have not considered as degrading the formation of noble products out of base materials and have rather been disposed to celebrate this as one of the exemplifications of wisdom and contrivance in the arrangements of Nature.[6]

In other words, Mill noted the ability of associationism to dissolve the complexity of faculty into simpler psychological elements and, in the process, to confront intuitionist philosophy. So did Maudsley. Maudsley considered that associationism explained why certain universal notions emerged and this explanation excluded the idea of the existence of *a priori* philosophical ideas.

Because all men have a common nature and because the nature by which all men are surrounded is the same, there are developed certain ideas which have a universal application but they are nowise independent of experience; on the contrary, the universality of their character is owing to the very fact that in every experience they are implicitly suggested or prompted, so that they finally become fixed as endowments in the acquired nature or organisation of the nervous centres.[7]

This characteristic— the dissolution of complexity into simple constitutents — was easily fitted within evolutionism. In the 1850s and 1860s this was precisely the turn that associationism was taking. In the *Descent* Darwin drew upon this tradition. It solved innumerable problems for him. It made the process of sophistication of faculty a process over time. In the *Descent* the complexity of man in relation to his primate ancestors stood in the same way as in associationism, the fully developed personality stood in relation to the simpler elements of which it was composed. Therefore the acquisition of complexity in human characteristics could be stretched over an evolutionary canvas. History could, therefore, be seen as the given time during which this complexity was developed. It also had another function. If, argued Darwin, 'Differences of this kind between the highest man and the lowest savage are connected by the finest graduations. It is possible that they might pass and be developed into each other.'[8]

The solution to these problems was not, of course, Darwin's invention. Spencer had struggled himself with the notion of discontinuity in the *Principles of Psychology* in 1855, though in a highly elliptical fashion. 'Even were the highest physical life', he had written, 'thus *absolutely* distinguished from psychical life . . . it would still be true that psychical life, in its earlier and lower phases, is not thus distinguished but arises only in the course of that progression by which life attains to more perfect forms.'[9] Therefore Spencer did have some justification on his part — although he was notoriously sensitive on points of precedence — when he wrote to Edward Youmans in 1871 that,

Since the publication of Darwin's *Descent of Man* there has been a great sensation about the theory of development of Mind . . . all having reference to the question of Mental Evolution and all proceeding on the supposition that it is Darwin's own hypothesis. As no one says a word in rectification, and as Darwin himself has not indicated the fact that the *Principles of Psychology* was published five years before the *Origin of Species* I am obliged to gently indicate this myself.[10]

In fact Darwin's contemporaries — or at least some of them — saw the derivative character of the *Descent* and its relation to associationist psychology. Frederick Pollock for example, in the first edition of the periodical *Mind* in 1876, was able to place both Spencer and Darwin within a pre-existing tradition. He described them both as part of the tradition of British empirical philosophy. It was only by the second generation of Darwinians that the independent importance of the tradition had become obscured by the prestige of Darwin's name.

The importance of the theory of mental evolution was this. By taking the complexity of faculty as pre-given, by using history as the means by which this complexity was achieved and by describing this process as one of graduated evolution, Darwin fell into the trap of progressive developmentalism. In other parts of his work he explicitly denounced progressive evolution. 'But yet there is no necessary tendency in the simple animal to become complicated'[14] he writes in the *Notebooks*. But it was possible to find denunciations of it and at the same time exemplifications of it in his work — specifically that on human evolution. Darwin could of course argue that man was a unique case and his history must be explained as the process by which, as Spencer put it, 'life attains to more and more perfect forms'.[12] However, this involved a radical departure from his position as set out elsewhere. It produced a comparative psychology of man in which the 'savage races' must be found a place on the lower rungs of graduated evolution — how else, from Darwin's point of view, could the evolution of mental faculty be explained 'naturally'? It was not, in Darwin's case, primarily a political or social ploy to construct an evolutionary psychology of this kind although it was not insignificant that it had social and political sanction. However, it was possible to find ethnocentrism and assumptions of European superiority supported by quite different intellectual positions. If they were there in Darwin's work it was partly because of the heuristic trap into which he fell and to whose brink he was led by the construction of the developmentalism he had adopted in the *Descent*.

II

Darwin's work on social evolution also showed his debt to the cultural and intellectual milieu of the 1860s. The *Descent* was written in the context of the debate which took place in the late sixties, in the British Association and elsewhere, between the progressionists and the degenerationists.[13] Briefly, it revolved around the question of whether 'savage' tribes could be treated as exemplifying the past of 'modern' social systems or whether they were, in Archbishop Whately's terms, the barbaric remnants of previous civilisations. It also had another dimension. Not only did it have implications for the study of anthropology, it also reflected upon the possibility of any scientific treatment of the development of social organisation. The degenerationist theory tended to imply that civilisation had been the result of the Divine Instructor. Man was put on earth with the rudiments of a social organisation according to a Divine plan. Some parts of humanity who had been given the benefits of civilisation had dissipated them through moral decline or intellectual debility.

At this point the proponents of the 'new' anthropology and Darwinism crossed paths. The implications of the *Origin* was that man had an historical beginning and that, in addition, his subsequent development was the result of natural forces. It meant that one must presume as Lubbock argued in his address to the British Association in 1867 that the first men and women made the first society. Anthropology was about the explanation of how and in what manner this occurred.

The epoch in which the first recognisably human and social being emerged remained to be explained. Those anthropologists who felt they had already found or discovered some near relation to the first society undoubtedly benefited from the impact of the *Origin*. Whether their characterisation was correct or not they had assumed some sort of historical beginning for civilisation and Darwin's theory seemed to uphold the necessity for this.

It became, therefore, common to integrate social evolutionary theory with a prologue drawn from Darwinism and this was particularly so after the publication of the *Descent*. In the textbook *Anthropology* published by Tylor in 1881, for example, the first few chapters were given over to a summary of the evidence in Huxley's *Man's Place in Nature* and the *Descent* about the genealogy of man with the primates and the suggestions in the *Descent* on tool-making among animals and the behaviour of primates in the herd. Afterwards Tylor reproduced his own particular theory of evolutionary development. However the theory of descent from the primate became the first step in the development which he went on to describe. Darwin did not inspire the theory of social evolution but he did provide it with the means by which the most inaccessible part of it — the direct emergence of man from the state of nature — might be explained.

However, Darwin's debt to the anthropologists was as great as their debt to him. He had been asked to provide a comprehensive theory of human development. The anthropologists provided him with one element of it, the first social state of man. In the *Descent* he accepted their premise that the past of man could be constructed by the investigation of existing primitive societies and that the lineage of contemporary society was indicated by the survivals within them of 'primitive' ideas and customs.

The evidence that all civilised nations are the descendants of barbarians, on the one side, consists of, clear traces of their former low condition in still-existing customs, beliefs, language, etc.; and, on the other side, of proofs that savages are independently able to raise themselves a few steps in the scale of civilisation and have actually thus risen.[14]

Moreover,

It is hardly possible to read Mr McLennan's work and not admit that almost all civilised nations still retain some traces of such rude habits as the forcible capture of wives. What ancient nation, as the same author asks, can be named that was originally monogamous? The primitive idea of justice, as shown by the law of battle and other customs of which traces still remain, was likewise most rude. Many existing superstitions are the remnants of former false religious beliefs.[14]

Only in one area did he permit himself criticism of the anthropological establishment of that period. For the purposes of defending his notion of sexual selection, Darwin attacked, very apologetically, the notion of primitive promiscuity and those Darwin described as 'the three authors who have studied it most closely, namely, Mr Morgan, Mr McLennan and Sir J. Lubbock'.[14] Wallace and Darwin both made this criticism of the anthropology of their day and it was through the influence of the *Descent* on Westermarck, the anthropologist, and his relationship with Wallace that it finds its way into established anthropological theory around 1890.

Hence an alliance was forged between Darwin and the anthropologists but one based upon certain limited objectives. It did not extend to the details of the theory of natural selection nor even to the developmentalism in the *Descent*. As Tylor put it in the 'Introduction to the Second Edition' of *Primitive Culture* in 1873,

It may have struck some readers as an omission that in a work insisting so strenuously on a theory of development or evolution mention should scarcely be made of Mr Darwin and Mr Herbert Spencer . . . . This absence of particular reference is accounted for by the present work, arranged on its own lines, coming scarcely into contact of detail with the previous works of these eminent philosophers.[15]

Tylor saw the incompatibility between the developmentalism of the *Descent* and his own theory of human evolution. Darwin was confined by the structure of explanation in the *Descent* to the notion of a psychological ladder of evolution on which the primitive races occupied the lower rungs. Tylor, on the other hand, might see social development in this way but he believed in a theory of evolution which presupposed that human nature could be taken as a relatively constant factor throughout history. Although customs and knowledge might differ, this did not imply, in his opinion, that the bases of human nature were a fundamental evolutionary variable.

But although progressive developmentalism as applied to human psychology lost him wholehearted support in some quarters, it also won him friends. Both Leslie Stephen and Alfred Marshall described the post-Darwinian epoch as having been affected by a revolution in social thought based upon this aspect of Darwinian science. Economic science must, therefore, according to Marshall writing in the *Principles of Economics*, reconstitute itself upon the premise that 'economics, like biology, deals with a matter of which the inner nature and constitution, as well as the outer form is constantly changing'.[16] By this he meant that it must incorporate within it the notion of a developmentalist psychology. In Stephen's case the political reasons for embracing this aspect of developmentalism are much more explicit. It was fuel for his attack upon the politics of utilitarianism. According to Stephen, developmentalism rendered baseless the view that,

Slavery explains the whole divergence (between black and white races) and that a negro differs from a European only as a man in a black coat differs from one in a white.[17]

III

What role was played in Darwin's thinking on human evolution by society itself? First of all, as with many other human attributes, he found the social instinct in a primitive form in the animal. This allowed him, for example, to be highly critical of some of the current anthropology of his day which, in a sense, had deprived animals of any significant social organisation. According to these theories, they were simply the chaotic predecessors out of which human evolution developed the rudiments of order and structure. In this respect earlier interpretations of animal society were more advanced. They had recognised structure within those societies even if they had simply used this fact as evidence of Providential influence or hung a sermon about human society upon it. Darwin argued that social organisation was an important factor in the survival of many species and consequently made an appearance early and throughout nature.

Darwin, in addition, used the notion of social organisation for another purpose. He argued that a complex of other characteristics was built on the basis of the social instinct. In particular the moral and altruistic faculties which supported and upheld social organisation were developments of it.

> The following proposition seems to me in a high degree probable — namely, that any animal whatever, endowed with well-marked social instincts would inevitably acquire a moral sense or conscience, as soon as its intellectual powers had become as well developed or nearly as well developed, as in man.[18]

Hence moral behaviour depends for its existence upon another primary fact of development — social organisation. This also solved an additional problem, that of the utility of morality. For, as Darwin's critics maintained, the evolution of a character must, under Darwin's scheme of things, be related to its usefulness. Of what utility they asked was altruism since in many cases it might be destructive of the individual or even of the race itself? In a similar way Darwin was faced with the necessity of explaining the utility of the aesthetic sense. He did this by linking it to the function of sexual selection and hence to the reproduction of the species. Morality, in Darwin's interpretation, played an essential role in the functioning of societies. It was the means by which social solidarity — itself an essential pre-requisite for the survival of the race — was cemented and maintained. In Darwin's opinion, morality had a functional character. However, Darwin also argued that though the origins of morality lay in its functional character, the development of rationality meant that moral behaviour became less instinctive, less dependent on its naturalistic origins and more open to reflection and conscious choice as evolution progressed.

This notion, however, was not original to Darwin. Spencer was under considerable attack from the periodical press of the time for suggesting that morality arose from social utility. But Bagehot's articles of the sixties, in particular, strengthened this part of Darwin's argument. Bagehot's work underlined the notion that social organisation was at the forefront of human evolutionary development and that it was closely linked to the moral and ethical character of societies at different stages of development. This hypothesis gave rise to one of the more influential schools of social Darwinism, concentrating upon those elements in the *Descent* which asserted the genesis of moral codes in their social utility. This school was, in contrast to other social Darwinisms, to develop along highly organismic lines. For underlying the notion of social utility suggested by Darwin was the belief that the rationale of ethical codes was in the social solidarity they gave rise to; secondly, in the obligation they put upon the individual to serve the social organism as a whole. When W. K. Clifford claimed, for example, that,

Some remarks of Mr Darwin's (*Descent of Man*, part 1, Ch. 3) appeared to me to constitute a method of dealing with ethical problems bearing a close analogy to the methods which have been successful in all other practical questions.[19]

it was this aspect of Darwinism to which he was referring.

The price paid for this alliance with developmentalism was considerable. It affected, for example, Darwin's reactions to all the texts of the 1860s. He saw Galton's articles on the differential birth-rate between the professional and working classes and his warnings about the effect this would have on the level of the nation's intelligence largely in terms of the theory of evolution which he had developed. Some of the more obscure passages of the *Descent* can be disentangled if we read them as Darwin's reply to the degenerationist tone of Galton's work. The other aspects fell, as it were, by the wayside. In fact, in the grip of the system he had constructed, Darwin's use of material for the *Descent* was to sort and sift the information he collected into support for or opposition to his theory. Consequently the *Descent*, as well as having a derivative character, is confused, self-contradictory and obscure in places. In seeming recognition of this were the lapses which he makes into what is in fact a form of pragmatic history — always, as far as the social theorist is concerned, the harbinger of despair. 'Progress', he argued at one point, 'seems to depend on many concurrent favourable conditions, far too complex to be followed out', and he went on to describe the effect of climate from Buckle, of 'fixed' property vide Henry Maine and of the 'union of many families under a chief' in Bagehotian terms.[20]

More important, however, was the fact that Darwin remained essentially entrapped within the assumptions of the critics he set out to answer. The individual whose evolution he describes is essentially the same being delineated by his critics such as Mivart and the Duke of Argyll. The conceptions of his critics were built into the answers he offered them. Therefore they had already won the battle since they had set the limits within which an answer had to be found.

In many ways Darwin's notion of development could be described as a reassertion of the 'great chain of being'. But it was distinct in that it had the 1850s and 1860s stamped on it in terms of its incorporation of the 'new' psychology and anthropology. It also changed the terms in which the debate on social and human evolution was discussed. His contemporaries (even if they mistook much of what he said for an original contribution) were quite right in seeing his work as inaugurating a new epoch. By allying an important scientific discovery to a certain conglomeration of social and cultural ideas, Darwin moved discussions of social development in certain directions.

IV

In 1912, Hyndman of the Marxist Social Democratic Federation replied to Wallace's letter congratulating him on his seventieth birthday:

It was therefore specially gratifying to me that you should be so kind as to write such a very encouraging letter on the occasion of my seventieth birthday. I owe you sincere thanks for what you have said, though I may honestly feel that you over-praised what I have done.[21]

In 1910 the magazine *The Social Democrat* reminded its readers to 'never forget that Wallace shares the honours with Darwin for the discovery of the law of evolution — and we may proudly add, is a Socialist'.[22]

This claim was not a matter of thrusting attention upon the largely indifferent. On the half centenary of the *Origin*, Wallace wrote a commemorative piece for the socialist newspaper *The Clarion*. By the end of his life his opinions had become part of the mainstream of social protest which punctuated the year before the outbreak of war. In 1913 Wallace wrote to a correspondent about the industrial discontent of that year. In his view,

This (high wages) is a principle worth enforcing by a general strike. Nothing less will be effective — nothing less should be accepted.[23]

Wallace's intellectual life spans a half century. Early in his career his views were influenced by Owenite socialism. This was overlaid in the sixties and seventies by Spencerian radicalism — even to the extent of naming his first-born Herbert Spencer Wallace. The rift between Spencer and Wallace grew in the 1880s when Spencer increasingly devoted his polemical skills to anti-socialism. In contrast, Wallace took up the cause of land nationalisation and became one of the followers of Henry George, the American land reformer. His views on Spencer were increasingly soured by the latter's repudiation of the traditional radical liberal demand of the mid-sixties — the nationalisation of land. In the 1890s and 1900s Wallace witnessed the emergence of political socialism and, without making too much distinction between the various strands, occasionally offered his skills in its defence. He also helped to educate a generation of socialists in the principles of natural selection.

However, Wallace's influence on the social content of Darwinian evolution is more complex. Wallace contributed a number of works on nineteenth-century anthropology. He helped shape the perceptions of a generation of intellectuals to the meaning of natural selection as applied to man. He also gives an important insight into the intellectual context in which the theory of natural selection was formulated.

Wallace spent 1848–52 in the Amazon Basin and 1854–62 in the Malay Archipelago. There he passed considerable time with the native peoples of the area and observed their life and customs. His observations both of the natural history and anthropology of these regions were published— *Travels on the Amazon* in 1853 and *The Malay Archipelago* in 1869. His work is primarily descriptive but it played a small part in the development of anthropology. Anthropologists of the mid-nineteenth century lent heavily on travellers' reminiscences of 'primitive' peoples. Both Darwin and Lubbock mention Wallace's anthropological observations.

Wallace's anthropology played only a small part in the anthropology of the nineteenth century because it was rather off-focus in relation to issues then current in the study of 'primitive' peoples. Kinship and religion were the predominant obsessions of nineteenth-century anthropology — a fact which reveals a great deal about Victorian culture. In spite of his subsequent interest in intellectual and moral evolution, Wallace's attention in these books was directed chiefly at the material basis of the culture of the 'primitive' peoples. Wallace's accounts of the Amazon and Malay Archipelago are filled with descriptions of methods of fishing and cultivation, the tools and utensils he saw, the mode of house-building and the degree of technical expertise he encountered. Many of the questions he set himself to discover relate to issues of this kind — why diet and culitvation differed between tribes, how, in a culture without iron nails, houses could be built.

Nonetheless a structure of explanation emerged from these early observations of primitive peoples. Wallace's anthropology closely paralleled his interest in natural ecology. He asked very similar questions about the peoples he encountered to those he asked about other organic forms. These were questions of how well a region could support a population; what were the natural checks on its expansion; the relationship between subsistence and size of population. His other preoccupation was with the geographical distribution of peoples. He put much greater emphasis than Darwin upon the role of geographical isolation in the evolution of species and varieties. Similarly he attributed many of the human racial differences in the Malay area to geographical isolation. Wallace was also interested in the effect on human evolution of that other major plank of natural ecology— migration. He spent some time in the classification of the languages of the Malay region partly for the clues they might reveal about the migration patterns of the peoples in the area. Perhaps for this reason Wallace thought more favourably than many of his contemporaries of the geographical determinism of Buckle's *History of Civilisation in England* (1858).

There was an even more interesting point which emerged in his travel writings. Wallace, like Darwin, put considerable emphasis upon Malthus's theory of population outrunning subsistence as a stimulus to evolutionary change. However, from time to time in his travel writings what struck him was the smallness of the human population in regions of abundant natural resources. In other words he asked himself the question of why, in certain areas, there were so few people.

During my residence among the Hill Dyaks I was much struck by the apparent absence of those causes which are generally supposed to check the increase of population although there were plain indications of stationary or but slowly increasing numbers. The conditions most favourable to a rapid increase of population are an abundance of food, a healthy climate, and early marriage. Here these conditions all exist. . . . Why then, we must inquire has not a greater population been produced?[24]

Wallace considered that among the Hill Dyaks population increase was limited by the extent of hard physical labour among women of child-bearing age. In other places he argued that the growth of population depended not only on the division of labour between the sexes but on the division of labour in general. Though he never formulated a general explanation of social evolution and population growth based on the advance in the material arts and social organisation, he showed himself in these early works to be preoccupied with these questions. The slow laborious toil he observed among some of the peoples of the Amazon region was,

principally the result of everybody doing everything for himself, slowly and with much unnecessary labour, instead of occupying himself with one kind of industry and exchanging its produce for the articles he requires . . . the consequence is that his work produces but sixpence a week, and he is therefore all his life earning a scanty supply of clothing in a country where food may be had almost for nothing.[24]

Wallace saw social evolution in these early writings as a process in which

Population will then certainly increase more rapidly, improved systems of agriculture and some division of labour will become necessary in order to provide the means of existence and a more complicated social state will take the place of the simple conditions of society which now obtain among them.[24]

Division of labour increased wealth but it also produced, according to Wallace, the conflicts which disfigured his own society.

These opinions have a particular interest for social theory. Wallace's discussion of human population growth recall strongly many of the early socialists' objections to Malthus. Wallace probably became acquainted with these in that period of his life, from 1837 onwards, in which he frequented the 'Hall of Science' off Tottenham Court Road in London. He described this as,

a kind of club or mechanics' institute for advanced thinkers among workmen, and especially for the followers of Robert Owen, the founder of the Socialist movement in England.[25]

It is, at least, a possibility that from this tradition Wallace acquired the view that, as far as human societies were concerned, there were *social* not simply natural limits to population increase.

These writings show Wallace, like Darwin, treating social organisation as an adaption. But unlike Darwin Wallace was more keenly aware of the notion of imperfect social adaption. Both Wallace and Darwin saw in maladaption in natural ecology a dynamic of change. Wallace's observations of badly adapted species in the forests of the Amazon destroyed for him the notion of perfect equilibrium in nature. According to Wallace the idea that every animal has a food particular to it and the physical organisation best adapted to secure that food had 'more of imagination than fact'.[26] The principle of maladaption he also saw to be true of societies — his own included. In the case of social evolution Darwin largely abandoned this idea for the belief in evolution as the emergence of superior individual adaptive faculties. The *Descent* virtually subjected human social evolution to the idea of increasing perfection — according to certain nineteenth-century ideas of what this meant. In contrast Wallace concentrated at this stage less on individual faculties and more on social organisation. His reactions to the native peoples he met show, on the whole, the absence of the kind of expectations a developmentalist approach would have aroused. Nor was he particularly impressed by the efficiency of many of the societies he encountered and apart from its productive power, of the viability and efficiency of his own.

If Wallace had developed and extended observations of this kind, a very original explanation of human social evolution might have emerged — one that put much greater emphasis on subsistence, population distribution and social structure. However, Wallace never succeeded in shaking off the notion that improvement in the material arts must be due to some intellectual advance or technological invention. Therefore these arguments were never sustained although one could argue that they found practical expression in his renewed interest in land reform and socialism later in his life. So that Wallace lapsed, in the 1860s, from an interest in social structure into making intellectual and mental development the cause of social change. The intellectual milieu around these questions, even in their most resolutely 'scientific' and 'materialist' forms, in, for example, Buckle and Mill, still emphasised the role of psychical and mental factors. Thus when in 1864 Wallace set out the theory of natural selection as applied to man in a paper before the Anthropological Society, it returned very strongly to conventional ideas of the importance of mental factors in evolution.

To the debate which was to rage in anthropology in the mid decades of the nineteenth century around Darwinism, Wallace added his own contribution in 1864 in a paper given to the Anthropological Society on 'The Origin of the Human Races and the Antiquity of Man Deduced from the Theory of Natural Selection'. This was important for Darwin in the construction of the *Descent*. It was a major blow in defence of his theory both of the animal descent of man and the unity of the human races which followed on that proposition. The other aphorisms in it crept into the *Descent,* namely the belief that moral and mental evolution had replaced to a large extent physical evolution. According to Wallace,

Tribes in which such mental and moral qualities were predominant, would therefore have an advantage in the struggle for existence over other tribes in which they were less developed, would live and maintain their numbers while the others would decrease and finally succumb.[27]

In addition, Wallace's belief in the total suspension at some date of natural selection in man emerged briefly in this paper.

To this extent, Wallace contributed some of the ideas upon which chapters in the *Descent* were built. But thereafter a rift began to grow up between Wallace and Darwin about what Wallace called the inadequacy of Natural Selection to account for the emergence of the mental and moral characteristics of the human species. Wallace says of this disagreement,

This view was first intimated in the last sentence of my paper . . . in 1864 and more fully in the last chapter of my essays in 1870.[28]

Darwin, however, first became aware of them through Wallace's review of the second edition of Lyell's *Antiquity of Man* in the Academy in 1869, in private correspondence with Wallace on the subject of sexual selection, and in Wallace's 'The Limits of Natural Selection as Applied to Man' which appeared in the *Quarterly Review* of 1869.

In these articles various evidences were adduced to show,

the existence of some power, distinct from that which has guided the development of the lower animals through their every-varying forms of being.[29]

The inference was that a superior intelligence had guided the development of man in a definite direction, and for a special purpose. In other words, the production of the moral and mental qualities of man could not be accounted for in the way in which his physical constitution could. There was, according to Wallace, a qualitative leap in evolutionary development at the point at which man emerged as a fully sentient creature. This could only be accounted for by supernatural causes, about the operation of which Wallace, however, remained rather vague.

It could be argued that there was an element in Wallace which saw that the complexity of 'savage' life could not be covered by an evolutionary theory of the kind that Darwin was propounding in the *Descent*. Loren Eiseley also takes this view but carries it much further. He sees Wallace as resisting the implications that Darwin appeared to be creating in the *Descent* of a hierarchy of races with the primitive peoples at the lower end of the scale and therefore nearer the animals. We can therefore, as he does, record Wallace's many tributes to the ingenuity and high moral tone of the primitive peoples and his insistence that they were closer to 'modern' man than to the beast.

But a view of this kind would contain only a grain of truth. Listen to Wallace:

The intellectual and moral, as well as the more physical, qualities of the European are superior . . . (and these) enable him when in contact with the savage man, to conquer in the struggle for existence and to increase at his expense. . . .[29]

Further,

And is it not the fact that in all ages, and in every quarter of the globe, the inhabitants of temperate have been superior to those of hotter countries? All the great invasions and displacements of races have been from North to South, rather than the reverse; and we have no record of there ever having existed, any more than there exists day, a solitary instance of an indigenous inter-tropical civilisation.[29]

Was Wallace therefore succumbing to a religious sentiment? Certainly he dabbled in spiritualism at this time. But in 1866 he wrote to Darwin to advise him to change the name 'natural selection' to Spencer's term 'survival of the fittest' because of accusations from Darwin's critics of 'blindness in your not seeing that Natural Selection requires the constant watching of an intelligent "chooser" ' and that 'thought and direction are essential to the action of Natural Selection'.[30]

There were basic inconsistencies in Wallace's views — a belief in some superior intelligence guiding human evolution and at the same time a strong resistance to attempts to invoke it in explanation of the development of other species. There was also Wallace's insistence on European superiority combined with his praise of the moral and mental capacity of the 'savage'. Later, however, Wallace repudiated his idea of the innate inferiority of primitive peoples. He also returned, via his interest in social reform, to placing the problem of material subsistence and social organisation at the centre of his view of human society.

On the whole therefore, it is probably correct to attribute Wallace's self-contradictory position to some basic intimation that human history required something more complex than a theory of graduated psychological and social evolution leading from prehistoric man to the present day. It is perfectly consistent with the idea of the importance of social organisation in evolution to infer that major qualitative change could be generated by variation in it, and that the 'gap' in evolution between man and primate may have been the product of a social rather than a spiritual revolution. The pity is that Wallace, mesmerised by notions of intellectual development, could not escape, any more than Darwin, the ideological limits of his time. Therefore he saw 'gaps' and 'qualitative' leaps still as the product of 'psychic' factors.

However, Wallace, whatever the inconsistencies, always denied that human evolution represented a simple graduation. In 1881 Wallace reviewed Tylor's *Anthropology* which set out a historical reconstruction based in part on the *Descent*. Wallace's comment on it was again an insistence on the inadequacy of developmentalism.

The extreme remoteness of the origin of man is also shown by the facts, that neither the size nor the form of the cranium of prehistoric races shows any inferiority to those of existing savages; while the approximate equality of their mental powers is shown by the ingenious construction of weapons and implements and the artistic talent we find developed at a period when the reindeer and the mammoth inhabited the south of France. It had been argued that the inferiority of the early implements shows mental inferiority but this is palpably illogical. Did Stephenson's, first rude locomotive — the *Rocket* — show less mind in its constructor than the highly finished products of our modern workshops?[31]

This view *was* related to his conception of the intellectual and moral qualities of 'primitive' peoples. In the debate on the notion of evolution by gradual progression which followed the British Association meeting of 1869 Wallace took up Lubbock's 'proofs' of the evolution of the moral sense. Wallace certainly agreed that there had been intellectual progress. How else could the improvement in the material arts be accounted for? Nonetheless he considered that Lubbock's case on *moral* evolution had not been proved.

He had met with savage tribes destitute of the arts of life and low in intellect, but possessed of a wonderfully delicate sense of right and wrong in morals. How did they get that sense? (Hear! Hear!) He had met some savages who would refuse to do an action which they thought would infringe on the rights of others and had refused to answer questions lest they tell a lie. . . . If they represented the original state of man, how came the moral sense to have grown and the other faculties not to have grown (Applause) . . . morals were hardly a scientific question; but he still thought that on its determination depended the perception of the true state of early man, they ought not to conclude that because man had advanced in the arts of life that therefore he had advanced in morals.[32]

These views ranged Wallace on the side of the Duke of Argyll and Stafford Northcote. The latter, who was to become leader of the Conservative Party in the House of Commons, had nominated Huxley as President of the British Association with evident reluctance and distaste. These were among the people who 'objected in a chorus of voices' to Huxley's intervention in this debate and applauded Wallace's. It was an ironical position for Wallace that his allies should be drawn from this social and political strata.

What were these debates about? Argyll and Northcote had no sentimental affection for the 'primitive' peoples. On most issues of negro emancipation and Colonialism Darwin and Huxley had a more liberal position. For Argyll and Northcote the issue was the defence of revealed religion. The savage's moral sense was merely proof of the divine creation of man and divine creation had also led to the dominance of the Church of England, the landed aristocracy and the British constitution. Similarly, they suspected that their opponents' determination to put the divine origin of man and society under scrutiny was, in many cases, a way of putting the religious and social authority of their own social group under a critical reappraisal. In many ways, as we shall see in the next chapter, they were right.

III DARWINISM AND LIBERALISM

THE strong influence which social Darwinism had upon conservative thought has tended to obscure its probably equal
influence over liberalism. The decade which followed the
publication of the *Origin of Species* in 1859 was one in which
radical liberalism was strong among an important section of
intellectuals and social Darwinism, which tended to take its
character from the prevailing social and intellectual atmosphere, reflected this. J. S. Mill had recently published his
defence of intellectual freedom in *On Liberty* (1859). Liberal
intellectuals in the 1860s were involved in issues — University
Test Acts, the Governor Eyre controversy — which acted as a
catalyst in the liberalisation of British intellectual and social
life.[1] The questions of the day were, on the whole, framed in
terms of liberal radicalism — how far could aristocratic influence in British life be diminished? To what extent could
economic life be freed from state control? What were the limits
on political and intellectual freedom?

Certainly Darwinism was used from the beginning as a
defence of 'laissez faire' capitalism. However, it was also used
in the attack on the remaining areas of special social and
political privilege in British society. The first important
attempts at elaborating its relationship to social issues were, in
fact, defences of ordered constitutional progress and more
social equality. This reflected the characteristic preoccupations
of the liberal intellectuals of the 1860s.

Both W. R. Greg, the Manchester economist, and Francis
Galton, the founder of eugenics, used the idea that society
could be divided between the 'fit' and the 'unfit' to attack
aristocratic privilege and landed property as well as to moral-
ise about the potentially disastrous effects on evolutionary
progress of a high birth rate among the working classes. They
considered the higher social stratum as impediment to
evolutionary development. Aristocracy, by awarding social

35

status for reasons of birth rather than achievement, protected
the idle and unproductive in society. Landed property mean-
while discouraged the exercise of economic initiative and
blocked social mobility upwards.[2]

Darwin himself reflected on the adverse effects on race
improvement of certain aspects of the social system. In a letter
to Hooker he wrote,

I have sometimes speculated on this subject; primogeniture is dreadfully
opposed to selection; suppose the first-born bull was necessarily made by
each farmer the begetter of the stock![3]

Galton summed up this alliance between race improvement
and the attack on privilege when he described in *Hereditary
Genius* (1869) the kind of social system most conducive to an
improvement in the character of the population. This was one
in which,

society was not costly; where incomes were chiefly derived from professional sources and not much through inheritance; where every lad had a
chance of showing his abilities, and, if highly gifted, was enabled to achieve a
first-class education and entrance into professional life, by the liberal help of
the exhibitions and scholarships he had obtained in his early youth; where
marriage was held in as high honour as in ancient Jewish times; where the
pride of race was encouraged (of course I do not refer to the nonsensical
sentiment of the present day which goes under that name); where the weak
could find a welcome and a refuge in celibate monastries and sisterhoods and
lastly where the better sort of emigrants and refugees from other lands were
invited and welcomed and their descendants naturalised.[4]

This was not the only attraction that Darwinism had for the
liberal intelligentsia of the 1860s. The reception of the *Origin*
and the controversy which followed seemed to raise major
issues of intellectual and social freedom. They followed the
debates between Huxley and Wilberforce with great
enthusiasm. On the one hand they saw Huxley as the pro-
tagonist of intellectual freedom and objective enquiry. On the
other, Wilberforce seemed to represent religious obscurant-
ism, an appeal to constituted authority and social prejudice.

However, the battle against these had been accelerated
rather than initiated by Darwin. Part of the importance of
Darwinism was that it was able to be integrated within existing controversies about the role of thought and science in the

social system. Many liberal political positions in mid-nineteenth century sought the support of philosophical and scientific ideas. In particular, liberal radicals found themselves locked in combat with intuitionism in philosophy. D. G. Ritchie in the 1880s described this battle in these terms:

thus Liberalism came to be identified with the criticism and removal of repressive laws and institutions, and an intellectual basis for such a policy was naturally found in a philosophy of critical analysis. It was in the same spirit that Locke, the father of English Empiricism, criticised the doctrine of innate ideas and the doctrine of the divine right of kings. And this alliance between Empiricism in philosophy and liberalism in politics continued with exceptions to the time of John Stuart Mill. . . .[5]

John Stuart Mill had described intuitionism as the 'intellectual support of false doctrines and bad institutions'.[6] By intuitionism he meant the doctrine that 'truths external to the mind may be known by intuition or consciousness independently of observation and experience'.[6] In particular, it was a philosophy which could be used to defend the idea of the divine inspiration of thought and of religious belief. In addition it suggested that certain institutions — religious and political — were not the result of historical or social development but the outcome of ideas planted in the mind by God and, for this reason, largely unchangeable.

As well as being a bastion of religious metaphysics, Mill suggested that intuitionism played a political role.

Now the difference between these two schools of philosophy, that of Intuition and that of Experience and Association, is not a mere matter of abstract speculation; it is full of practical consequences and lies at the foundation of all the greatest differences of practical opinion in an age of progress. The practical reformer has continually to demand that changes be made in things which are supported by powerful and widely-spread feelings . . . and it is often an indispensable part of his argument to show, how these powerful feelings had their origin. . . .[6]

In other words, among the obstacles to the progressive improvements in society that liberals demanded was a philosophy which argued for the immutability of human nature and the divine origin of human ideas.

Mill was not a social evolutionist but he did find a means of arguing against theories of human nature in the associationist philosophy of his day. This argued that human faculties were derived from the reaction between the human organism and the environment. This was the philosophy of Experience and Association to which he referred. Ritchie, as we have seen, described this as an alliance between liberalism and empiricism. But Darwinism pushed arguments based on this philosophical position a stage further. In the *Descent of Man* Darwin produced a series of explanations of the origin of morals, intellect and social feeling by reference to the relationship between the faculties of the individual and its environment. Frederick Pollock placed Darwin and Spencer side by side as far as this was concerned. What they both

teach us is to extend to the race as a whole the process and the conceptions which the English school of empirical philosophy has already applied with great success, as far as it went, to the individual.[7]

The success of the theory of natural selection had justified, in the opinion of some, the 'progressive' philosophy of human development.

As Frederick Pollock said,

if Mr Darwin is right they (the intuitionists) must be wrong. The Intuitionist denies that moral sentiment can be accounted for by the materials given in the experience of the individual.[7]

The impact of this debate was considerable. Pollock referred to the fact that,

For two or three years the knot of Cambridge friends of whom Clifford was the leading spirit were carried away by a wave of Darwinian enthusiasm; we seemed to ride triumphant upon an ocean of new life and boundless possibilities. Natural selection was the master key of the universe; we expected it to solve all riddles and reconcile all contradictions. Among other things it was to give us a new system of ethics combining the exactness of the utilitarian with the poetical ideals of the transcendentalist.[7]

Three aspects of Darwinism came together in a litany of liberal belief. The first was the defence of Darwin's right to propagate the theory of natural selection and to imply that man's origin ought to be examined from a strictly scientific basis. This right was defended by J. S. Mill although he, himself, was not convinced that Darwin had successfully demonstrated the operation of natural selection. The second, which we shall examine shortly, was the use of Darwinian analogies to defend political liberalisation and to attack privilege. The third was the explanation of the origin of man's faculties by reference to the relation between sensation and environment. These three elements intertwined. Out of a combination of the second and third came one major strand of social Darwinism closely associated with the group referred to by Pollock as 'carried away on a wave of Darwinian enthusiasm'. This was the attempt to give ethics a scientific foundation by explaining their development as the social equivalent to the process of natural selection. W. K. Clifford set out this theory in a series of articles published posthumously in the 1870s. Pollock reiterated it in an article in the first issue of *Mind* in 1876 and Leslie Stephen finally synthesised it in *The Science of Ethics* in 1882.

The main idea, as set out by Clifford, was that,

Society is an organism, and man in society part of an organism; according to this definition, in so far as some portion of the nature of man is what it is for the sake of the whole-society. Now conscience is such a portion of the nature of man and its function is the preservation of society in the struggle for existence.[8]

In other words the origin of morals was to be explained by their usefulness in binding society together. This was certainly derived from Darwinism in so far as in the *Descent of Man* Darwin had used the idea of 'utility' to explain aspects of social behaviour. Clifford acknowledged that,

Some remarks of Mr Darwin (*Descent of Man*, Part 1, Chapter 3) appeared to me to constitute a method of dealing with ethical problems bearing a close analogy to the methods which have been successful in all other practical questions.[8]

These remarks of Darwin could be summarised in the following passage from the *Descent*.

Turning now to the social and moral faculties. In order that primeval men, or the ape-like progenitors of man, should have become social, they must have acquired the same instinctive feelings, which impel other animals to live in a body; and they no doubt exhibited the same general disposition. They would have felt uneasy when separated from their comrades, for whom they would have felt some degree of love; they would have warned each other of danger, and have given mutual aid in attack or defence. All this implies some degree of sympathy, fidelity, and courage. Such social qualities, the paramount importance of which to the lower animals is disputed by no one, were no doubt acquired by the progenitors of man in a similar manner, namely, through natural selection, aided by inherited habit. When two tribes of primeval man, living in the same country, came into competition, if the one tribe included (other circumstances being equal) a greater number of courageous, sympathetic and faithful members, who were always ready to warn each other of danger, to aid and defend each other, this tribe would without doubt succeed best and conquer the other. Let it be borne in mind how all-important in the never-ceasing wars of savages, fidelity and courage must be. The advantage which disciplined soldiers have over undisciplined hordes follows chiefly from the confidence which each man feels in his comrades. Obedience, as Mr Bagehot has well shown, is of the highest value, for any form of government is better than none. Selfish and contentious people will not cohere, and without coherence nothing can be effected. A tribe possessing the above qualities in a high degree would spread and be victorious over other tribes: but in the course of time it would, judging from all past history, be in its turn overcome by some other and still more highly endowed tribe. Thus the social and moral qualities would tend slowly to advance and be diffused throughout the world.[9]

Although the emergence of a social theory was an important aspect of these developments in Clifford's 'circle', so too was the anti-religious bias. Stephen explained in a letter to B. J. Norton in 1877 that *The Science of Ethics* upon which he was then working had the aim of 'to put what may be called the derivative or scientific theory of morality so as to meet objections in their newest shape and to show how morality is independent of theology'.[10] Stephen continued in the 1880s and 1890s to put forward these views to ethical societies. In this way they became part of the secularist tradition in Britain. But this theory, as well as a defence of the non-religious ethic, was also a sociology.

Theories of this sort explored organicism — a biological analogy much older than Darwinism — to explain social phenomena. Organicism was used to emphasise the importance of social and moral bonds in keeping society together. This had been a highly conservative notion in the early nineteenth century and its adoption by liberals involved a number of modifications. Conservative organicism had suggested two things: the necessity of hierarchy in society (a controlling and directing centre) and the interdependence of the various parts of society. The latter idea implied that major social engineering was impossible without disturbing the whole and this might lead to the morbidity or death of the social organism. Liberal organicism certainly took over these ideas but it also transformed the organic analogy. If Darwin had shown that natural organic evolution occurred, then why not social evolution? Social evolution would have to take account of the 'vital tissue' binding society together. This excluded violent disruption. But it was nonetheless possible to envisage major social change paralleling natural change.

In addition, other biological ideas argued against the notion of the all-importance of a central social authority. The controlling functions of the body were not all centred in one organ but dispersed throughout the organism. The centre was, indeed, more dependent upon the sub-systems than previous biologists had realised. Moreover important adaptations and variations were initiated within these organic sub-systems. These might lead to the slow, gradual modification of the organic whole. Above all, when transferring these ideas to society, liberal Darwinists adapted them to Darwin's theory of moral evolution. What bound the social organism together was not authority and political sovereignty, but moral and social sentiment. This required the diffusion of democratic and civil sentiment throughout the society.

The complex uses to which Stephen put the organic analogy in *The Science of Ethics* owes much to Spencer. But it also recalls modern functionalism. Stephen believed that sentiments and values held society together. In *The Science of Ethics* he described a range of social institutions from the family to the political club. All, he argued, produced a distinctive system of sentiments and moral beliefs. Naturally in order that society should survive it required the integration of these systems within an overall social ethic. He considered that the family acted as the chief means for transmission of this overall ethic. This process was never completely successful so that, in practice, there was always the chance of conflict between different institutional values and the demands of social order in general.

Talcott Parsons in *The Structure of Social Action* (1937) outlined a theory of social action which he saw as in revolt against nineteenth-century conceptions, particularly the utilitarian theory of action. But he shared with most later nineteenth-century theorists a belief in the generation of social order through the ideas of obligation planted in man's head by social institutions. He also, like Stephen, found that this idea could be integrated with notions of system and function derived from organicism. But a distinction between *The Structure of Social Action* and the late nineteenth-century thinkers arises over the question of rationality. Liberal Darwinists insisted on the importance of reason in evolution. This was partly because of the analogy they borrowed from Darwinism between organic and social 'variations'.

II

In a series of articles in the late sixties, Bagehot harnessed biology to his theory of politics. Like many late nineteenth-century liberals he was concerned with social order. His observations of France from 1848 to 1851 had given him the opinion that, 'The first duty of society is the preservation of society . . . . To keep up this system we must sacrifice everything. Parliaments, liberty, leading articles, essays, eloquence, all are good but they are secondary.'[11] These views, which contrasted with those of many liberals in 1848, led Bagehot to emphasise the importance of sentiment in society such as the sentiment which bound the masses to the symbols of monarchy. In spite of this Bagehot's *Physics and Politics* (1872) was a defence of liberal democracy by use of a Darwinian analogy. It emphasised cultural rather than individual selection and sought to prove that the institutions and practice of liberal democracy was the guarantee of evolutionary progress.

It also used the notion of social utility. All social institutions were fitted to their historical epoch. Moreover, because the early stages of human society were dominated by the crude struggle for existence between tribes and nations then institutions which ensured the greatest fighting strength and unified purpose were best suited to these epochs. Thus early society was dominated by political despotism and militarism. Moral cohesion underlined and made more effective this social unity.

As the political and institutional bonds loosened, moral solidarity became more important. But they must loosen, argued Bagehot, if any evolution towards superior social forms was to take place. Evolution, according to Bagehot, was the loosening of those bonds which held men together by institutional and political means. It was the substitution of military despotism by constitutional government.

This emphasis on evolution as a movement towards constitutional government was justified, in Bagehot's view, by a Darwinian precept. Both Bagehot and Stephen applied the notion of selection of variations to human evolution. These variations were essentially *social* variations. They behaved in an analogous way to the variations which occurred in nature. The 'fittest' survived and changed the direction of social evolution. According to Stephen, 'a modification due to the social factor would be for our purposes in the same position as an organic modification'.[12] Institutions, forms of government, cultural changes, in general, evolved out of these social variations.

Whilst biologists puzzled about the origin of organic variation, Stephen and Bagehot believed they knew the causes of social variation. A social variation could be a new technology, a cultural advance or a new political institution. It could also be an intellectual discovery. Stephen's opinion that ethics and religion were not necessarily interdependent fell into this category. It was, according to Stephen, a progressive social improvement. It reduced the role of superstition and of religious authority in society and, therefore, would lead to a surer foundation for moral behaviour. Ultimately all these innovations were the product of the intellect, the free play of ideas and the evolution of rationality.

A political conclusion could be drawn from this. Only those societies which allowed a degree of intellectual freedom would give rise to social variations — the material out of which social evolution occurred. Thus the best political institutions from the point of view of social evolution, were liberal ones. Bagehot applied this idea to his analysis of ancient and 'primitive' societies. These 'primitive' societies met the demands of social order and defence but if their institutions became too rigid and conservative to allow innovation, debate and change, this would halt their social and political development. They would atrophy and, perhaps, become extinct.

Ideas of this sort were used to defend the liberal view of the world as a group of peacefully competing states. Physical struggle for survival was characteristic only of early society. In later stages, cultural competition replaced physical. The characteristic of an idea or invention was that it could be copied. Bagehot, in particular, believed the natural imitativeness of man was a major source of social evolution. So long as fashion favoured the superior in society then nothing but good could come from the imitation of this ideal by social inferiors. In one sense Bagehot gave the monarchy the role of delineating in its behaviour the desirable social ideal.

Stephen, too, believed that evolution took place by the peaceful transmission of ideas between peoples and nations. This again, depended upon the development of certain conditions — lack of xenophobia, openness to change, friendly relations between states. This world order paralleled that of Cobden's version of free and equal states. But the free trade Stephen proposed was in ideas not goods. There was one qualification. To be admitted to this world, a people must have reached a certain level of intellectual and social development. Stephen's views did not apply to the 'primitive' or 'savage' whom he considered were too intellectually, morally and socially inferior to take part in this process.

These analogies between the idea as a 'variation' and natural selection have a counterpart in philosophies of scientific progress. Karl Popper has described the growth of knowledge as, 'the result of a process closely resembling what Darwin called "natural selection": that is, the *natural selection of hypotheses*'.[13] Popper claimed that the 'conscious and systematic criticism of our theories' provides the environment in which scientific hypotheses survive or become extinct.

Nineteenth-century pragmatism produced a similar sort of explanation of scientific and intellectual progress although it was much more committed to positivism than Popper's philosophy. Thus the tests theories had to pass were experimental and practical as well as theoretical. Although Stephen and Bagehot were primarily concerned with social theory, the two strands — the philosophical and social — cross. Michael

Polyani, for example, in the twentieth century put forward the hypothesis that the critical environment, in which theories had to survive, was provided by the democratic and liberal charac-ter of scientific institutions and, further, that this was threatened or protected by the degree to which the political institutions of society as a whole were liberal and democra-tic.[14]

III

So far liberal social Darwinists seemed to have ensured the importance of reason in evolution and of morality indepen-dent of religious ethics. However, they had also to demons-trate that rationality, even if it destroyed religious authority, was not subversive of morality and social order in general. They did this by sociologising ethics.

Liberal social Darwinists strongly argued that reason and morality were linked, or, at least, not opposed. It was reason-able to be moral partly because it could be shown to be socially useful. This was not a residue of utilitarianism. Social utility, mediated through Darwinism, came to mean something quite different from the calculus of self-interests. It meant an histori-cal and sociological investigation of the role of morals. This implied there was no absolute standard of morals applicable in every historical society but that moral codes were relative. It also made the needs of the social organism rather than the consultation of individual self-interest the means by which the social utility of morals was measured. Neither did it depend, as Parsons argues, on assuming rational calculation as the source of social behaviour. Many nineteenth-century theorists, as he correctly points out, devolved the source of social action upon heredity thus avoiding discussion of motive and intentions — rational or otherwise — altogether. But others, though they believed it was possible and even necessary to rationally *see* the social good, believed that this social good quite often involved a sacrifice of self interest. It was the self interest of the social organism not the individual that social behaviour served. The rationality of behaviour was a 'higher' *social* rationality not an individual one. In this respect modern functionalism is based on precisely the same assumption. Nonetheless liberal Darwinians still insisted that even if the end of social action was not individual interest but the good of society it was still possible and necessary to see what that end was and to discern the means that served it.

The Cambridge moral philosopher Henry Sidgwick wondered whether this tradition of ethical discussion would be finally swallowed up by sociology. But this never happened, partly because it never abandoned the search for a transcendental moral ideal. Thus it transformed Darwinian evolution into the realisation through history of a moral archetype. Stephen constantly used terms such as 'ever perfecting type' in describing organic evolution. Similarly he searched for the basic seeds of some eternal moral order in historical societies. This philosophical bent became even more powerful under the impact of T. H. Green.[15]

However, the tradition was important. In some hands, the notion that it is socially useful to be moral and a course of action recommended to us by reason and nature led to theories of a rather shallow character. But to understand why they occurred, it is important to see the conditions in which they were formulated. They arose when the traditional philosophical supports of social order and moral influence were being undermined by certain scientific developments. This was taking place at a time of rapid social change. In these circumstances to argue that knowledge of science, history and evolution actually awakened moral awareness and social responsibility seemed important.

Nonetheless this strand of 'Darwinian' thought fed a stream of vulgar evolutionary optimism. A number of texts of this sort emerged. Some, like Henry Drummond's *Ascent of Man* (1894), were an attempt at Christian apologetics. If the theory of moral evolution was correct then it seemed, in Drummond's opinion, to be the best possible evidence for divine intervention in human affairs. Another section of works attempted a more 'scientifically' inspired account of the origin of various moral faculties. Alexander Sutherland, for example, in *The Origin and Growth of the Moral Instinct* in 1898 — a book which had some influence on Hobhouse — attempted to trace a genealogy of moral ideas using Darwinian methods. Sutherland once again returned to Darwin.

If the name of Charles Darwin but rarely occurs in these pages it is not that
they owe little to his influence. On the contrary, full half of the book is a
detailed expansion of the fourth and fifth chapters of his *'Descent of Man'*.[16]

These works followed closely the pattern revealed in Sutherland. They looked for the origin of morality in some instinct
which had emerged in early human or in late animal evolution.
From it they traced subsequent moral evolution as a sophistication of this initial instinct. Sutherland discovered the original basis of morality, not as Darwin did in the social instinct
but in the parental. The social instinct was, in fact, an extension
of the parental. Parents who cared for their offspring preserved them in contrast to the careless or improvident parent.
Therefore those who inherited this instinct survived and those
who lacked it died out.

There were many similar demonstrations, very few of them
showing much originality or departing from this basic structure. But the popularity of them demonstrated an important
aspect of British culture — a constant search for a scientific
basis for moral and social obligation. It also remained an
important stream of thought within liberalism.

Green's moral philosophy had two effects on this tradition.
Green was also concerned with the re-interpretation of liberalism, but deriving inspiration from Hegel not Darwin. However his disciples frequently attempted to unite his ideas with
the language of biological evolution. They were drawn in two
directions. Some like Samuel Alexander used organicism but
removed the sociological component from it. He was basically
more concerned to biologise philosophical language. Thus he
tried to find the biological equivalents for categories of obligation, the definition of 'good' and 'bad' acts. Others such as
D. G. Ritchie were deeply concerned with the social implications of philosophy. But Ritchie, for example, defended social
reform and collectivism by using the Hegelian dialectic and
the language of biological evolution to suggest the possibility
of a society which would transcend the conflict between individualism and collectivism.[17] In other words Green's influence
was to philosophise the debate and, in doing so, some of the
explicitly sociological concerns of Clifford, Pollock and
Stephen were dropped. Thus *The Science of Ethics* became a

relatively neglected text of the late nineteenth century in spite of the fact that Stephen's is reported to have held a high opinion of its value. 'I believe that in one sense he liked it better . . . than any of his other books. It showed, so he thought, that if he had not "scattered himself too much" he might have accomplished a good piece of scientific exposition.'[18]

However, this sociological strand reappeared in Hobhouse's work. L. T. Hobhouse, writing at the turn of the century, derived much from this tradition. He also altered its direction in a number of important ways. Whilst Stephen and Bagehot had been largely content with arguing their case for the marriage of reason and morality by analogy, Hobhouse attempted to give this evolutionary theory a firmer biological foundation. Hobhouse's lessons in physiology from J.S. Haldane and his articles on animal intelligence were part of a search for the actual historical moment at which both human intelligence and moral feeling emerged. His major works were attempts to provide the theory of rational and moral development with a scientific preamble.[19] Thus, although he was against Spencer's doctrine of biological laws operating in human society, he nonetheless considered that,

I could never accept the view that the whole work of science was of secondary importance, that it could go on constructing the world as it chose, but that, whatever its results, a metaphysical analysis would also be able to interpret the entire scientific scheme on its own lines. Doubtless metaphysical analysis and scientific specialism have each its sphere, but they cannot maintain an attitude of mutual indifference to the end . . . . In this respect Mr Spencer, whatever the defects of his method, seemed to me to have been justly inspired.[19]

Hobhouse's insistence on a scientific basis for morals reveals his strong links with the original anti-intuitionist school. Mill believed there was a 'natural hostility between (anti-intuitionism) and a philosophy which discourages the explanation of feelings and moral facts by circumstances and association'.[20] Hobhouse similarly required an explanation of human faculty in terms of natural laws and rejected a metaphysics of moral theory. Like Bagehot, Clifford, Stephen and Pollock, he turned to Darwinism to provide him with one.

Consequently, in spite of his warnings against certain

aspects of biological social theory, Hobhouse was strongly influenced by biology. The curriculum he produced for students at the London School of Economics included,

Physical, Psychological and Social Physical conditions (of society). Theory of Organic Evolution. Heredity. Psychological Conditions. The study of animal behaviour . . . . Social Morphology as the basis of Social Evolution.[21]

His reading lists included J. A. Thomson on *Heredity* (1908), C. Lloyd Morgan on *Animal Behaviour* (1900), R. H. Lock on *Variation, Heredity and Evolution* (1906) and S. Herbert on *The First Principles of Heredity* (1910).[21]

According to Hobhouse, 'Darwin in a couple of chapters in the *Descent of Man* made a valuable contribution to the explanation of the ethical side of mental development.'[22] Hobhouse's objection was therefore not to social Darwinism as such but to versions of it popular at the turn of the century. These, by their emphasis on heredity, implied that moral and rational choice was of less importance in evolution than Hobhouse and others of this tradition considered. That is why Hobhouse returned to an examination of the actual process of evolution. The objective was to show how the operation of reason and moral feeling could actually be observed in the evolution of human psychology. Hobhouse whilst he, in these crucial respects, took up the liberal tradition, also departed from it. He attempted to accommodate a measure of collectivism and social reform in liberalism. In doing so he moved the analogies familiar to Stephen, Pollock and others in directions which were unacceptable to them.

In contrast to the 'new' liberalism Clifford, Stephen and Bagehot would not accept that the precepts of moral evolution and organic solidarity led in the direction of either social reform or collectivism. On the contrary, they believed that these ideas justified economic competition. It was true that competitiveness must always be subordinated to social solidarity. Thus, according to Alfred Marshall,

. . . the struggle for existence causes in the long run those races of men to survive in which the individual is most willing to sacrifice himself for the benefit of his environment, and which are consequently the best adapted *collectively* to make use of their environment.[23]

However, the competitive ethic was still essential in society. From it arose economic innovation, and these economic innovations were among the variations out of which evolutionary progress was made. A great deal of social life was taken up by the 'struggle for existence' in the crude sense. Differences in capacity meant that some would succeed and others fail. Stephen criticised T. H. Huxley for his view that moral evolution had somehow suspended the struggle for existence. On the contrary Stephen believed that,

the more moral the race, the more harmonious and the better organised, the better it is fitted for holding its own. But if this be admitted, we must also admit that the change is not that it has ceased to struggle, but that it struggles by different means.[24]

Leslie Stephen also denied that morality enjoined on society the importance of providing social welfare. He argued, for example, that there was a class of apparently moral actions with disastrous social consequences. Into this category fell the excesses of the philanthropic spirit. Indiscriminate charity whilst apparently 'moral' was highly ambiguous as far as its social value was concerned.

Charity, you say, is a virtue; charity increases beggary, and so far tends to produce a feebler population; therefore a moral quality clearly tends to diminish the vigour of a nation. The answer is, of course, obvious, and I am confident that Professor Huxley would so far agree with me. It is that all charity which fosters a degraded class is therefore immoral. The 'fanatical individualism' of today has its weaknesses but in this matter it seems to me that we see the weakness of the not less fanatical 'collectivism'.[24]

Similarly Bernard Bosanquet refused to adapt his belief in the importance of moral evolution to accommodate the movement towards collectivism. In fact, he argued that what was important about the role of morality in society was not the obligation to carry out social reform, which some had argued was implied by the idea, but the obligation to follow the precepts of the Charity Organisation Society and combat poverty by producing a moral improvement among the poor.

Now broadly speaking the co-operative individual, as demanded by civilised life, can only be produced in the family . . . and the test and engine of his production is the peculiar form of moral responsibility, supported by law and covering both material and moral incidents, which the family implies.[25]

This meant that social improvement must be brought about by the moral regeneration of the poor, and the means to do this were, in Bosanquet's opinion, to save them from the degrading effects of state subvention.

As Stephen put it,

sobriety and prudence among the lowest classes might be improved . . . . The consequences of such a change would, I suspect, be incomparably greater than the consequences of whole systems of laws regulating the hours of labour and whole armies of official inspectors.[26]

These liberals had on their side the strong argument that morality implied choice. This included the choices made about the welfare of one's family; to save or to spend; to work or to be idle; to think of the future or only to consider the present. If the state reduced these choices, it reduced the sphere of moral action. But evolution could only take place by increasing moral choice. Interference by example or education was justified since to choose one must know what choices existed. But interference by a state assuming responsibility for these decisions was a negation of morality.

It could be argued that the divisions between the 'new' and the old liberalism hinged on attitudes which should be taken towards the rise of an independent working-class politics. But it could not be argued that the 'new' liberalism invented the notion of either organicism or moral evolution to justify collectivism. These were firmly part of a tradition which had grown up prior to the 1880s and 1890s. The old liberalism also offered a view of the relationship between the classes and a theory of their integration within society. But the means it suggested were not those of concession to the demand for social reform.

The theory of moral evolution and organic solidarity suggested that a scientific basis existed for the containment of working-class demands. It undercut political theories based on individualism and self-interest. To some liberals a working-class politics based on these notions would lead to violence and social instability. Instead, this view recast society in a way which emphasised community and de-emphasised conflict. But this did not necessarily mean a programme of social reform. It suggested a programme for instructing the working class in their moral duties, one of which was to moderate their political and social demands and to improve themselves rather than society. It could be argued that though the language in which these sentiments had come to be expressed were Darwinian, the character of the sentiments had deep roots in the tradition of nineteenth-century middle-class radicalism.

The first generation of political liberals — the grandfathers and uncles of Stephen and others — had been evangelicals. The second generation had — on the whole — lost their faith but, it could be argued, had rediscovered a version of it through the adaptation of Darwinism to a theory of moral evolution. Darwinism had not only secularised ethics, it had apparently ensured their survival by showing their rational and necessary character. Evolution gave values the transcendental character which Divine authority had once done — just as Pollock had predicted. Natural selection applied to social evolution suggested that morality was not only desirable but necessary. It also, as adapted by Stephen and Bagehot, suggested the possibility of steady moral improvement. Moreover, like the early evangelicalism, the values which were important in social evolution were those which had a bearing on public behaviour, on social responsibility and on the relationship between classes.[27]

Moral evolutionism did not, therefore, imply social reform or collectivism. Moreover, although its precepts came to be popular among British socialists, it did not imply equality. It was, in its own way, as hierarchical as those versions of social Darwinism which emphasised economic competition. Instead of economic 'fitness' intellectual fitness became the measure of evolutionary value. In particular, this view of evolution defended the intellectual in politics. In part, it was for this reason that it retained its popularity — at least among intellectuals. Intellectuals were agents of social progress, providing social variations by their ingenuity.

This brings out the links between Stephen's social Darwinism and some aspects of J. S. Mill's thought. Stephen had criticised Mill's theory of human nature from an evolutionary perspective. However, although Mill had argued that the case for evolution was not strong he nonetheless considered that 'there really is one social element which is thus predominant and almost paramount among the agents of social progression'.[28] This was 'the speculative faculties of mankind including the nature of the speculative beliefs'. Mill considered

It would be a great error . . . to assert that speculation, intellectual activity, the pursuit of truth is among the more powerful propensities of human nature . . . . But notwithstanding the relative weakness of this principle among other sociological agents, its influence is the main determining cause of social progress . . . .[28]

The social Darwinism of Stephen, Bagehot and Bosanquet implied that there was a moral as well as an intellectual elite. There were, in Bosanquet's view, those who were morally 'fit' and the morally 'unfit'. Evolutionary progress demanded either that the latter should receive moral instruction or that they should be brought under the control of their moral superiors. In either event it gave those who claimed certain moral knowledge a particularly important social role. In a sense the role of Evangelicalism was reaffirmed by these means.

Social Darwinist versions of liberalism represented the two sides of liberal politics. On the one hand they provided reasons for intellectual and political freedom and argued the dysgenic character of many of the social and political constraints existing in mid-nineteenth century Britain. They defended the possibility of an ethics without religion. They used naturalistic explanation of the origin of man's faculties against the protagonists of the notion of an unchanging and pre-ordained 'human nature'. On the other they justified hierarchy, moral superiority and social order. Later on proponents of the 'new' liberalism which advocated measures of social reform designed to bring the working classes within liberal politics, increasingly appropriated the language of social Darwinism. But it was classic liberalism which first attempted this. By the time the 'new' liberalism arrived on the scene, the social concerns of liberalism had shifted from the reconstitution and defence of freedom to the question of major reforms in the social system.

IV INDIVIDUALISM AND COLLECTIVISM

THE 1880's saw a change in the political atmosphere in which social Darwinism was discussed. Beatrice Webb, in her autobiography, described it in these words:

There were, in fact, in the eighties and nineties two controversies raging in periodicals and books, and giving rise to perpetual argument within my own circle of relations and acquaintances: on the one hand the meaning of the poverty of masses of men; and on the other, the practicability or desirability of political and industrial democracy as a means of redressing the grievances of the majority of people. Was the poverty of the many a necessary condition of the wealth of the nation and of its progress in civilisation? And if the bulk of the people were to remain poor and uneducated, was it desirable, was it even safe to entrust them with the weapon of trade unionism, and, through the ballot box, with making and controlling the government of Great Britain with its enormous wealth and its far-flung dominions?'.[1]

In other words, the 'social problem' had come to the forefront of nineteenth-century political consciousness. This was partly the result of increase in working-class radicalism in the 1880s. In 1881 the Marxist Social Democratic Federation was formed and in 1884 the Fabian Society came into existence to act as propagandists for social reform. This took place against a background of serious unemployment riots in London 1886–7 and the growth of trade unionism among the unskilled and semi-skilled. The end of the decade saw two strikes — that of the match girls led by Annie Besant— and the major one in the docks in 1889, both of which focussed attention on questions of social conditions and pay.

Intellectually these events produced several responses. Henry George, the American social reformer, toured Britain advocating schemes of land redistribution to solve the problem of poverty and unemployment. Joseph Chamberlain campaigned on an unofficial liberal platform which included measures of social welfare which he described as the ransom property must pay in return for security. The pay and conditions of the working class were discussed at a Conference on Industrial Remuneration in 1885. Investigators like Charles Booth began a systematic enquiry into the social conditions of the poor. In other words, as Beatrice Webb says, the question of socialism, state action to alleviate poverty and social welfare were increasingly the subject of discussion and controversy.

It is disputable to what extent this had practical political effects, but the intellectual ones were considerable. The 1880s saw a debate in progress between the advocates of some form of state intervention in questions of social welfare and its opponents. The first, often termed 'collectivists', included a spectrum of political opinion from liberal advocates of social welfare to full-blooded socialists. On the other side were those who opposed the extension of state intervention or who regarded themselves as the intellectual combatants of the socialist 'menace'. Before long social Darwinism was involved in these debates on both sides and the interpretations given to Darwinism in this debate continued to exercise influence well into the twentieth century.

In the 1870s social Darwinism shifted towards an emphasis upon moral, cultural and intellectual evolution. But this view of evolution, although popular among liberal intellectuals, never succeeded in dispelling other quite opposed theories of the operation of natural selection in society. The most influential of these was the use of Darwinism to defend competitive individualism in society and its economic corollary — laissez-faire capitalism.

Darwinism appeared to many as a means of justifying the attempt to gain an advantage over one's fellow men in a competitive race for economic success. In a humorous aside to Lyell, Darwin referred to versions of this when he wrote,

> I have received, in a Manchester newspaper, rather a good squib, showing that I have proved 'might is right', and therefore that Napoleon is right and every cheating tradesman is also right.[2]

Hobhouse was also referring to this view of social behaviour when he wrote:

> In the middle of the 'Eighties' when the writer was first studying philosophy the biological theory of evolution was already very generally accepted and

the philosophical extension of the theory by Mr Herbert Spencer was, in academic circles, in the hey-dey of its influence.[3]

Hobhouse called this a materialist system by which he meant, in this context, a theory of social behaviour which attributed to men and women no other motive than that of their well-being.

As we have seen, liberals accepted a degree of competitive individualism in society, but they also felt that social behaviour could not be reduced to vulgar self-interest. Their theory of society was based, on the contrary, on notions of altruism and regard for others. Economics was regarded by them as merely a sub-system within society and its character had to be modified to take into consideration the rules of social order. Social behaviour based purely on economic individualism was attributed to the breakdown of evolutionary progress rather than regarded as its natural product.

In these circumstances Spencer's use of Darwinism to defend competitive individualism in the articles he published in the '*Contemporary Review*' in 1884 came as a considerable shock.[4] Many liberals were aware that Darwinism had been used in this way in political and social argument but this was the first time this view of evolution had been given a philosophical justification by so eminent a person.

Spencer's answer to the questions outlined by Beatrice Webb was that poverty could only be alleviated by state intervention at the cost of social progress. The reason was that Spencer attributed to economic competition the same role which Darwinism had given natural selection. Economic competition weeded out the 'fit' from the 'unfit', the economic failure from the success. This implied that laissez-faire was the best condition under which economic competition and hence social evolution could take place. It implied also that there was some form of natural acquisitiveness in man.

In fact, Spencer's sociology was much more than an expression of economic individualism. He had, after all, developed a theory of moral evolution and emphasised organic solidarity in societies. Nonetheless, within these limits it is also true that he had treated society as a conglomeration of individual wills and choices which — because of structural links between aspects of the social system — somehow coalesced. His organicism was compatible with individualism. But was it compatible with the notion that moral evolution took place? His articles of 1884 seemed to describe individual behaviour in terms of competitiveness? Spencer was, in fact, sensitive to accusations that he had ignored moral sentiment in human social behaviour. In a reply to the criticisms that Huxley made of him for this emphasis in his articles, his answer was that he did not dissent from the view that nature was non-moral and that human society was moral. He agreed that society had made progress to the point at which war and struggle had been superseded. He claimed that he merely differed in his estimates of the advantages to be gained from social reform. This did not imply callousness on his part.

Because I hold that the struggle for existence and the survival of the fittest should be allowed to go on in Society, subject to those restraints which are involved by preventing each man from interfering with the sphere of action of another, and should not be mitigated by governmental agency, he, along with many others, ran away with the notion that (my belief was that) they should not be mitigated at all. I regard voluntary benevolence as adequate to achieve all those mitigations that are proper and needful.[4]

He was, he claimed, merely arguing against 'collectivism' and there was no reason why — as many liberals agreed — collectivism was the necessary end of moral evolution. In a sense Spencer's arguments sharpened the problem. The plethora of defences of vulgar self-interest in the 1880s alienated some from Spencer's defence of individualism. But a section of liberal opinion was also alerted to the use of moral evolution to justify welfare reforms and, after Spencer's intervention in the debate, they made — as Bosenquet did — careful attempts to show how their view that altruism was the basis of social evolution should not be confused with pleas for social reform by the state.

Out of Spencer's attack on 'collectivism' a fierce debate on the relationship between Darwinism and social reform arose. Patrick Geddes, then lecturer in zoology in the School of Medicine in Edinburgh, launched an attack on the individualism of classical political economy. Geddes's dislike of contemporary economic theory was largely based on a similar revulsion to that of Ruskin. He disliked a political economy based on a calculus of self-interest; it seemed to symbolise the dissolution of moral and social ties between men that contemporary industrial society had brought about. At the same time Geddes wanted to demonstrate that 'even on the most sternly biological grounds so far from a scientific basis for economic deduction being furnished by the "iron law of competition" . . . (its law) is the accurate converse of this — the golden role of sympathy and synergy'.[5] Like most of Spencer's critics Geddes refused to surrender the authority of Darwinism to the individualists. Unlike Ruskin, Geddes's attack on contemporary industrial society was based not on the rejection of a science of political economy but on its redefinition.

Darwin had:

not only revolutionised modern biology, and with it our views of the origin, nature and destiny of man, but has shed new and brilliant light upon the special sciences which concern him, anthropology, philogy, psychology and ethics — the economist alone remains behind, and although long ago armed with the purely biological ideas of competition and co-operation, delays to modernise his theories by aid of the new learning and treats them as if they were independent of such general conceptions and of struggle for existence, of functional differentiation and change, of polymorphism and the like of which they are really special cases.[5]

Geddes's refutation of classical political economy was based on a similar use of the organic analogy to that of Leslie Stephen. Leslie Stephen's *Science of Ethics* (1882) argued that a society, like an organism, is bound by structural links and these links are reinforced by the common acceptance of certain moral values by society. Few social theorists after him gave the organic analogy a more sophisticated treatment. Geddes was able, however, to 'biologise' the notion much more. He gave a biological name 'polymorphism' to structural differentiation and attempted to show its operation in nature and its increasing importance in the evolutionary development of the species. He also insisted in assimilating all aspects of economic life to biological function. Economics was 'but one special case of the vast problem of the modification of organism by the environment'.[5]

Like many social Darwinist texts of the 1880s Geddes's work was partly aimed at a defence of the role of social reform and state intervention. It was only a partial defence because Geddes envisaged much social reform as the product of voluntary rather than state agencies. But he did insist upon the necessity of planning the urban environment and mitigating the effects of it by conscious intervention. The justification for this was based on his conception of the social organism. The social organism like the natural one was threatened from within by disease. The city to Geddes was the main area of social morbidity. From it spread all the debilitation and weakness of society in general. The cure of urban problems he saw in terms of the treatment of an organic disease which if left untouched would be fatal to the organism itself. In his description of the function of social science Geddes used medical metaphors. Sociologists were like doctors attempting to cure certain ills which had arisen in the body politic. This was an image frequently repeated in this period, especially since a great many early sociologists were recruited into the social sciences from various forms of medical practice and biological research. This was true of Geddes, C. H. Myers, C. G. Seligmann, W. H. R. Rivers and A. C. Haddon.

In 1923 Rivers argued against Morris Ginsberg's view that the analogy between the social and the natural organism was of limited value. His conviction that this was not the case rested partly on the belief that medicine might eventually produce laws or at least concepts of treatment applicable to social problems. Rivers was cautious about the possibility but he refused to exclude it.

There is little question that, backward as it may be, the medicine of the organism is more advanced than the medicine of society. While the one is now founded on definite principles and laws, the other is still in the stage of pure empiricism. Nevertheless, it would be dangerous to apply the medicine of the individual to the disorders of society until we know far more than at present of the laws which regulate the normal working of society. At the same time, I believe that the statesman and the politician would be largely assisted in obtaining this knowledge by the lessons to be learnt from the discipline which has as its subject the morbid state of the individual and in this lecture I propose to consider as fully as time allows some points of

resemblance between organism and society from the pathological standpoint.[6]

The concepts applied by Rivers of social medicine, prognosis
and cure pushed the organic analogy towards an explicit inter-
ventionism. But at the same time it did not yet imply a defence
of state intervention. It merely demanded that social problems
should be tackled and put forward the opinion that they could
not be left to individual effort or benevolence.

   In the 1880s D. G Ritchie harnessed some of the presuppositions of 'new' Liberalism to an attack on Spencer. Unlike
Geddes, Stephen and other protagonists of the organic analogy, there was a much more explicit commitment in Ritchie to
a defence of the state. This was by no means necessarily
implied by organicism. It was perfectly possible to talk of
organic links in society whilst assuming that the controlling
functions were also widely dispersed among the social organism. A social organism did not imply a concentration of control in some centre analogous to the head or brain. In fact,
Spencer's attack on collectivism was, as many of his critics
saw, based on the idea that power and self-direction is widely
diffused among the structures which make up the social or
natural organic form.

   Ritchie's defence of the state represented two aspects of his
thought, T. H. Green's attempt to detach liberalism from a
purely negative view of the state's functions and Ritchie's own
association with the Fabian Society. The advancement of
social reform was seen by the Fabians as essentially a capture of
the institutions of the state and an expansion of their functions.
The state was, in their opinion, potentially a benevolent
institution. According to Ritchie,

The history of progress is the record of a gradual dimunition of waste. The
lower the stage, the greater is the waste involved in the attainment of any
end. In the lower organisms, nature is reckless in her expenditure of life . . . .
When we come to human beings in society, the State is the chief instrument
by which waste is prevented . . . by freeing the individual from the necessity
of perpetual struggle for the mere conditions of life it [the state] can set free
individuality and so make culture possible.[7]

Ritchie's appropriation of the organic analogy was essentially a reinterpretation of Spencer's use of it. In contrast to Spencer, Ritchie argued that the social organism was more highly structured and that the integrity of this structure depended not only on the moral links but the encapsulating of this moral community within a controlling centre — the state. In contrast, Spencer's notion of organic form was essentially 'not to be compared with any noble animal . . . but that it belongs to an extremely low type. We are "bodies dispersed through an undifferentiated jelly". This, we suppose, represents the British citizen moving in his national fog'.[7]

T. H. Huxley's intervention in 1887 and in 1889 added another dimension to the debate. Huxley took up the already widely applied notion of evolution as the development of moral feeling and rationality. But, he argued, unlike many who had previously used this idea, that this must lead to a break in the continuum which was supposed to link man and nature. In other words, he argued in favour of some form of evolutionary development which operated by a crude struggle for existence. This, however, belonged to early stages of evolution.

This individual self-assertion was made impossible by the emergence of moral feeling. Once this occurred the nature of evolution changed. The only possibility of returning to the crude struggle for survival was if population increase got out of control. Like Ritchie this view was essentially a justification of certain areas of state intervention — in Huxley's case education. This state intervention was justified by the moral need men felt for the mitigation of ills and also made possible by the ability of men to reconstruct rationally their environment. This meant that social development was actually a frustration of natural law rather than its expression.

Huxley's intervention posed problems for all the protagonists in the controversy.[8] In the first place Ritchie's arguments were based on the idea of the continuum between man and nature. He wrote,

Regarded as events in time, the appearance of consciousness and the capacity for language . . . may be accounted for by natural selection. . . . i.e. they favoured in the struggle for existence those species which happened to possess them.[9]

Ritchie began, however, to make qualifications about the survival of natural selection in the 'higher' social stages. In 1894 he wrote

Every careful biologist (the qualification is unfortunately necessary) recognises that evolution is not identical with what we mean by progress. Even in the biological sphere, success in the struggle may be attained by degeneration as well as by advance. And, as to social evolution, has not Professor Huxley very well said that the being who survives a free fight proves his ability to survive a free fight but no other kind of ability? The rise of ethical ideas may be explained *historically* in terms of natural selection, but when these ideals have arisen they make social progress something different from mere organic evolution . . . .[9]

But this was not acceptable to Leslie Stephen who went to some lengths to refute Huxley's notion that natural selection ceased to operate once moral ideas had arisen. The whole burden of the introduction of moral and rational factors into evolution which had taken place in the 1870s was not in order to distinguish social evolution from natural evolution but to emphasise an analogous process. Thus Stephen certainly agreed that, 'An individualism which regards the cosmic process as equivalent simply to an internecine struggle of each against all must certainly fail to construct a satisfactory morality . . .'.[10] At the same time he maintained that the emergence of morality was merely a process by which the needs of the race were secured consciously by social and moral selection — a process which had already taken place unconsciously.

The rhetoric of social reform was influenced by this debate. In Hobhouse's work *The Labour Movement* (1893) the influence of it can be traced. The disciples of the school of social reform believed, he argued, that 'society is more than a mere aggregate of individuals, and see it as a living whole'.[11] Like a human organism the social organism must be kept healthy. Moreover they, the social reformers, were also proponents of natural selection.

Now we fully agree with the evolutionists in their main position. It is desirable that the fit should succeed and the unfit fail; we are ready to exclude the utterly unfit from society altogether by enclosing them in prison walls . . . .[11]

But Hobhouse went on to distinguish between those fit to

survive in 'the unregulated contest of individuals' and those
who are 'fittest morally to survive in a society of mutually
dependent human beings. And that the morally fittest shall
actually survive is the object of good social institutions'.[11]

But there was nothing implicit in organicism or moral
evolution which pushed it towards collectivism, nor did liberal individualism entirely succumb to the new language of
social reform. By the 1900s there was a re-emphasis of some
of the earlier Darwinian analogies which had emerged in
the 1870s. By means of these individualists re-asserted their
distrust of state intervention. In 1906, for example, C. W.
Saleeby, asked to give a series of lectures by the British Constitutional Association against the growth of collectivism,
restated the thesis that the origin of evolutionary progress lay
in social variation and the condition which favoured it was a
free, uncontrolled environment.

I stand here as a biologist and my objection to collectivism, for the present, is
a biological and philosophic objection . . . . The one final objection to the
trade union which says that a clever workman may not work faster nor an
energetic workman longer than his neighbour is that such a practice is
fundamentally opposed to natural selection.[12]

II

There were two strands in late nineteenth-century British
socialism. The first — the revolutionary tradition — was much
smaller and less influential than the second, the reformist. On
certain points about Darwinism, however, they stood
together. These were their support for Darwin and their insistence that the precepts of socialism were in accordance with or,
at least, not in contradiction to the laws of natural selection.
The second point was their hostility to Spencer and their
identification of individualism as the main intellectual enemy
to socialism. Within this consensus there were wide
divergences and also considerable emnity between the factions
of British socialism about the meaning of Darwinism. The
relation of socialism to Darwinism reveals not just one series
of connections but a whole network of meanings.

Before there existed a socialist movement in Britain in any
serious form, Marx and Engels extended a welcome to Dar-

winism. His work was, firstly, a blow against religion and, Marx claimed, against teleology in the natural sciences in general. What did Marx mean by this? He meant, primarily, that Darwin had dispelled the notion of the pre-ordained in nature. Random variation meant that the adaption of creatures to their environment was no longer evidence of the existence of design. This, as Marx saw, was a principle which could be used when all evidence of structure and function in the universe was adduced as an example of an intelligence at work. Secondly, Darwinism opened up the history of human faculty to scientific treatment. It therefore dispelled theological dogmatism about human faculty. Rapidly Darwin became part of the secularist movement. Edward Aveling, in particular, undertook to popularise Darwin to a wider public in order to advance the anti-religious cause.[13]

In addition, Darwin provided confirmation of Marx and Engels' view of scientific development. The funeral oration given by Engels at Marx's grave was an attempt to reaffirm Marx's importance as a thinker by linking him with general historical developments in the sciences. Marx and Engels retained a residual Hegelianism which influenced their view of the order and development of the sciences. As Hegelians they believed that ideas had their time. With this in mind, they sought in other spheres of intellectual development for echoes of the general principles they believed they had elucidated in history. Their delight at the *Origin* was natural for a number of characteristics of it could be linked with Marx's achievements; the notion of objective forces moving independently of 'will' or direction and the possibility of constructing a history by the application of principles of this sort. This seemed to them to be evidence of the general historical importance of the ideas Marx propounded. Their view of scientific development was, therefore, reminiscent of all those texts of philosophy in the nineteenth century which sought for the great moving principle of thought reflected in various branches of science. This included ideas ranging from Spencer's application of the notion of persistence of force to the social as well as natural world to Edward Clodd's idea that the sciences of the nineteenth century were all evolutionary.

As far as this idea was concerned Marx and Engels were

truly nineteenth-century thinkers. However, their conception of their relationship to Darwin was considerably more sophisticated than that of Spencer or Clodd and deserves a closer examination. In the first place there is no justification for the conclusion that they saw in natural selection a reflected image of the class struggle. On the contrary, they went to some lengths to dissociate themselves from any notion that the historical reconstruction of nature and society were based on the same principles. Marx described Lange's attempt to apply a general principle to both as a way of escaping the concrete analysis of societies at different stages of development and under different modes of production, an effort which he called 'swaggering, sham-scientific bombastic ignorance and intellectual laziness'.[14]

Engels did do some borrowing from Darwin or more often from the intellectual environment surrounding Darwin. In *The Transition from Ape to Man* (1876) he produced a theory of how the individual acquired faculties through the interplay between its senses and the environment. This was an exact intellectual replica of the model used by Darwin in the *Descent of Man*. Engels also incorporated this with use-inheritance in a similar way to Darwin. This reveals that he had, on the whole, learnt more from the milieu around Darwin than from the *Origin*.

The Russian Marxist Plekhanov in the 1890s used the theory of the evolutionary development of man to argue that,

In that notorious 'human nature', there is not a single feature which is not to be found in some other variety of animal and, therefore, there is absolutely no foundation for considering man to be some special being and separating him off into a special kind.[15]

This theme became a common litany of much European and British socialist writing on human evolution. Even before the *Origin* and the *Descent* a non-religious conception of human evolution based on a sensationalist psychology had developed among the British secularist and socialist circles. As Edward Royle shows, in the disturbed years of 1837–44 in Britain when radicalism in thought and deed were at its height, there were philosophical defences in these circles of the naturalistic character of human development.[16] Wallace from 1837 became acquainted with them and when he left on his travels to the Amazon carried with him a high opinion of one such putative attempt — *Vestiges of Creation* (1844) by the deist Robert Chambers. Darwin's notebooks, compiled in this same period, were redolent of the kinds of argument found in these— at that period— disreputable circles which accounts in some measure for his extreme caution in making his views public. The secularist and socialist tradition had, therefore, every reason to claim Darwin as their ally. Moreover the 'transformation' of the Victorian 'world view' he is accredited with having accomplished must be seen in conjunction with the older tradition — from which it could be argued Wallace emerged — and which was made respectable by the intellectuals of the 1850s and 1860s.

However there were a number of aspects of the theory of natural selection which worried Marx and Engels. The primary one was the apparent reliance Darwin placed upon Malthus. Malthus's notion that population would always outrun subsistence and that, therefore, no society could solve completely the problem of poverty was widely used in the nineteenth century as an argument against socialism. Darwin claimed to have been influenced by Malthus to the extent that the struggle for subsistence which was described in the *Essay on Population* led him to the realisation of the intensity of competition in nature.

However, Marx and Engels were aware that Darwinism was being associated with a kind of revised Malthusianism. Arguments were being adduced about the inevitability of poverty and the persistence of the competitive spirit in the social as well as the natural world. This was described in ironic terms by Engels as,

A transfer from society to living nature of Hobbes' doctrine of bellum omnium contra omnes and of the bourgeois-economic doctrine of competition together with Malthus's theory of population. When this conjuring trick has been performed (and I question its absolute permissibility) the same theories are transferred back again from organic nature to history and it is now claimed that their validity as eternal laws of human society has been proved. The puerility of this procedure is so obvious, that not a word need be said about it.[17]

Efforts have been made to deduce from this passage the idea

that Engels saw in Darwin's theory of natural selection a reflection of capitalist society. If this was the case, it would be a surprising fact. Both Marx and Engels regarded the bourgeois-economic doctrine, exemplified in this passage, as an ideological distortion of the objective reality of economic life under capitalism. They considered the categories of importance in understanding the economy — whatever role competitiveness played as a practical ideology of the capitalist class — were the productive relations of society and the corresponding class relationships. Individualism was not the 'secret' motive power of the economy whatever part it played in the philosophy and political economy of capitalism. Neither Marx nor Engels were prepared to jettison Darwinism because of the 'apparent' similarity between it and these ideological notions.

As far as identifying what kind of social interpretations would be given Darwinism, Engel's characterisation proved largely correct. Wallace struggled harder than his contemporaries in attempting to relate the idea of production to social evolution. Though his ideas, in some respects, owed more to Adam Smith than Robert Owen, they had the merit of treating production as something more than pure economic individualism. In contemporary sociobiology, only two notions of production play any substantial role. One is the idea of entrepreneurship (dominance) as increasing differential fertility; the other is war (aggression) as a means to increase wealth. Both are exact replications of 'bellum omnium contra omnes' and neither posits an economic relationship any more sophisticated than that of individual appropriation of an existing stock of goods.

Engels was aware of the way in which a world view was emerging which integrated natural selection with the slogans of competitive capitalism. The passage quoted was in a letter to the Russian socialist Lavrov in 1875 and its import was in fact advice on how to handle arguments of this kind. Engels told Lavrov to dispute the Malthusian antecedents given to natural selection and also to emphasise how production had irrevocably changed human evolution. He also advised him to combat the notion of individualism by arguing that nature provided as many examples of co-operation as it did of competitiveness. In other words, his advice was to engage in an ideological battle over the significance of Darwinism. But that the problem of human nature was of least significance in this battle is illustrated by Engels' criticism of the German socialist Duhring's attempt to sentimentalise nature by expunging competitiveness and struggle from Darwin. However, the most crucial area in the relationship between Marxism and Darwinism was the question of teleology and its relation to the social and natural sciences. Marx mentions this question once in his published letters (1861) but in *Anti-Duhring* (1878) Engels went into this question at great length. Herr Duhring's attack on Darwin's theory of evolution and his proposal for an alternative, Engels identified with a philosophy based on,

. . . the Hegelian 'inner purpose' — i.e. a purpose which is not imported into nature by some third party acting purposively, such as the wisdom of providence, but lies in the necessity of the thing itself — (this) constantly leads people who are not well versed in Philosophy to thoughtlessly ascribing to nature conscious and purposive activity . . . .[18]

This led to conceptions of natural evolution in which the idea of adaption was used to infer the existence of purposive activity or the realisation of thought and motive in organic forms. On the contrary argue Engels,

If therefore tree-frogs and leaf-eating insects are green, desert animals sandy-yellow . . . they have certainly not adapted these colours on purpose or in conformity with any ideas.[18]

What significance did this have for social theory? Duhring, Engels argued, had not only restored consciousness and insisted upoon the immutability of ideas (though not necessarily forms) in nature, he had done precisely the same thing for society. Motive and consciousness in both nature and society had these characteristic outcomes. Their importance rested on their influence over individual behaviour. This dissolved social relationships into the problem of individual motivation and consciousness. These motivations were reconstructed by the theorist through introspection. Engels called this, in relationship to social theory, a process by which the object consults its concept of itself and then proceeds to be 'measured by its image, the concept'.[18] In religious treatises this concept was God's idea of the world reflected in reality. In social theory it tended to be the individual's concept of his role and behaviour in society. This meant that social relationships disappeared as a constitutent of social investigation. Society became built around the social molecule 'by the multiplication of which the whole of society is to be built up'.[18] Moreover what the social molecule found in consulting his consciousness was,

For the most part, moral and juridical notions which are a more or less accurate expression (positive or negative, corroborative or antagonistic) of the social and political relations amidst which he lives; perhaps also ideas drawn from the literature on the subject; and, as a final possibility, some personal idiosyncracies.[18]

III

Engels' interpretation was in effect a good analysis of the tradition of social Darwinism, especially of the kinds of transformations of Darwin's view of the organic world necessary to accommodate this kind of social theory. It is interesting to observe Engels's insistence upon a social world independent of consciousness and upon the futility of making the social molecule a basis of social analysis, in the light of the recent rediscovery of a theory of 'human nature' in Marxism (however 'negative and antagonistic' towards existing social and political relations this is).

However, in Britain there was no serious socialist component in 'respectable' philosophy. Disreputable philosophy was still largely built around secularism and naturalism. Moreover, the social theories of Marxism only permeated a tiny section of socialist intelligentsia. The debate around Darwin and socialism was, therefore, almost wholly linked to attacks on religion, upon the notion of an immutable human nature and to using Darwinian analogies created by the liberal 'collectivist' tradition mediated by thinkers like Ritchie and Geddes.

Socialists were not altogether unprepared for the attack launched by Spencer in 1884. Lafargue reviewed his articles for the *Contemporary Review* in *Today*. He claimed that Spencer had misunderstood Darwin. The theory of natural selection did not imply an acquisitive human nature which social change would leave untouched.

It was generally presumed that Mr Spencer had understood the Darwinian theory of which he had volunteered to be the propounder. The anti-Socialist axiom cited above inclines us to think that the presumption was erroneous . . . . For, according to the evolution theory, the origins of animals, their habits and their instincts are not spontaneous growths . . . but have been evolved more or less gradually by the actions and reactions of their milieu. Different conditions of life must consequently produce different instincts and habits.[19]

But Spencer's use of Darwin also gave rise to a certain dichotomy in Socialist attitudes to Darwinism. A year later, in 1885, socialists celebrated the unveiling of Darwin's statue. According to *Justice,* organ of the Social Democratic Federation,

To us scientific Socialists whose sociological theories are based upon recorded facts of evolution and development, Darwin's work serves, as a sure foundation.[20]

On the other hand, at the end of the decade another contributor to the paper noted that,

Every discovery of science, every invention of mankind, has been seized upon by the bourgeoisie to delude and exploit the proletariat . . . . In a like manner the bourgeoisie accept the teachings of Malthus and pervert those of Darwin to bolster up the tottering fabric of society today and they steal from the armoury of the evolutionist weapons which they use in their own defence.[20]

British socialism based its precepts, largely, on the idea of moral evolution and a regenerated human nature. In their distaste for Spencer's individualism, they found liberal allies who were consciously offering the hand of friendship to socialism. Thus Alfred Marshall, in many ways the heir to mid-nineteenth century classical economic individualism, found a place in his social Darwinism for aphorisms of social and moral solidarity.

Similarly at the Conference of the British Association in 1885, Patrick Geddes seemed to offer biological interpretations of social evolution as a possible compromise between the poles of individualism and socialism. According to Geddes,

The biologist must side with the individualist against the socialist in recognising that man can never shake himself wholly free from the iron grip of nature, yet undiscouraged by this, since recognising the vast modificability of life through its surroundings, must yet encourage the socialist in every rational effort to subordinate the natural to artificial selection and raise the struggle onto the culture of existence.[21]

Geddes's offer of reconciliation to socialists was, however, limited. In 1886 in *The Claims of Labour* Geddes used the organic analogy which earlier he had deployed against Spencer to argue against socialism. Structural differentiation implied social authority was distributed unequally within the social organism. According to Geddes the cerebral functions were the most important. This meant that notions of the value of labour were romantic and misplaced. The aim of the social organism was the reproduction of social life and in this process the controlling and directing functions were as vital as those of labour. Moreover, the redistribution of wealth would, in Geddes's opinion, destroy the necessary functions performed by the capitalist entrepreneur.

Geddes was regarded with some suspicion by socialists. A. P. Hazell in *Justice* attacked him as 'the writer' (who) puts forward some special views of his own on Natural Science especially dilating upon the dependence of social upon physical and biological science, pitying the poor, ignorant unscientific Social Democrats'.[22] Apparently Geddes's contribution to socialist debate included an attack on the trade unions. His concept of social organic health required that they may be converted into something more closely approaching the medieval guild and that they should lose their rights to bargain on conditions and pay. But Geddes persisted in his efforts at reconciling socialism and his own version of social welfare. In 1895 he appended a foreword (with J. A. Thomson) to a book by the Belgian socialists Massart and Vandervelde.

The two Belgians drew a parallel between organic and social parasitism arguing that the health of both the natural and social organisms was endangered by it. Social parasites they defined as 'those who appropriate and consume without contribution to life by 1. Fraud or violence, 2. Exploitation of debauchery and the sexual passions, 3. useless public offices, 4. appropriation of the means of production'.[23] The political message was clear. Massart and Vandervelde equated capitalism with parasitism and advocated its forcible suppression. The only viable society was one composed of those who produce. From this kind of society both the rentier at one extreme of society and the lumpenproletariat at the other were to be excluded. Although it contrasted with Geddes's own view of social peace, he considered it to be a 'concrete instance of that practical union and unison of Biology and Sociology which has been so loudly proclaimed by the philosophers but so little carried into practice'.[33]

A much more influential book was *Socialism and Positive Science* (1894) by the Italian socialist Enrico Ferri who quite consciously tried to make socialism the heir to certain aspects of classical liberalism. For example, he accepted that human inequality existed and that only the 'fit' should survive. He also argued that it was the constitution of modern society which prevented this. Only in conditions of social equality — the abolition of inherited wealth and advantages — could this happen. This, Ferri claimed — much to Spencer's chagrin — made socialism a natural extension of Spencerian evolution.[24]

Ferri's book combined the combativeness characteristic of continental socialism with aspects that had greater appeal for the mainstream of the British Labour movement. For example, he believed the growth of population led to the struggle for existence but he considered that as society ascended the evolutionary scale the pressure of population grew less, partly as a product of technological advance. The law of organic solidarity took its place. Moreover, whilst accepting the efficacy of 'survival of the fittest' in evolutionary progress, Ferri also argued that the concept was relative. In a society of fools, the most foolish is the fittest. Socialism, by providing a society which was moral and rational, ensured that the fittest was also the 'best', something which could not be ensured by the crude struggle for existence.

These openings towards a less rigorous and more pacific theory of evolution made his book highly acceptable. Ramsay MacDonald quoted him and so, too, did the *Social Democrat*.

To the more radical social democrats Ferri seemed to offer a more acceptable definition of the 'fit'. In a free society in which opportunity would arise from the extinction of privilege it was by no means the case that the fit would be found among the upper classes.

If it be true that the theory of the survival of the fittest in nature is antagonistic to Socialism and that the most fitted to survive are those that procreate the most there is a seeming paradox in nature when applied to social conditions. We find in nature that individuals most procreative have the best chance of preserving the type in conflict with others and securing the most favourable basis of life. In the social order the reverse obtains. The types possessing the greatest economic power over their fellows rot and idle through luxury, debauchery and ease, while the workers increase their numbers freely. The reason for the non-extinction of the least fit is because of the creation of a force in society to repress any inclination of the masses to put in operation against the classes the natural law which operates in the lower forms of life. This force is called political power.[25]

Socialists in these exchanges showed their capacity for turning the metaphor of survival of the fittest to their own advantage, though they had fewer effective means for disseminating their view of Darwinism. On the whole, however, Ferri was quoted in so far as he merged with the existing tradition not of socialism but of the liberal radical view of evolution. It was Huxley, Ritchie, Geddes and Stephen's view of evolutionary development which had more influence in the mainstream of the socialist movement. This was committed to gradualism rather than revolution; to morality and rationality rather than political struggle; and to social harmony rather than class war.

They were also committed to Darwin. Keir Hardie, for example, in 1906 wrote, 'It was in the early eighties that what became known as Darwinism thus gave a fillip to the competitive system by appearing to stamp it with the sign manual of scientific approval.'[26] In contrast, Hardie argued that,

Darwin stated emphatically that 'those communities which included the greatest number of the most *sympathetic* members would flourish best', and in so stating he conceded the whole case for which the Socialist is contending. It is sympathetic association and not individualistic competition which makes for progress and the improvement of the race'.[26]

British socialism was suffused with moral evolutionism. The

basic argument of many socialists against individualism was a repetition of the liberal notion that altruism and sociability were the qualities which best aided men in the struggle for survival. Henry Drummond's *Ascent of Man* with its vision of universal harmony guaranteed by evolution fitted perfectly the conceptions of many socialists and was widely quoted. Keir Hardie even accepted the religious aspect of it. Hardie's view of evolution was filled with pietistic optimism to the extent that his formulation of evolution, as the secularists pointed out, verged on the edge of anti-Darwinian. For in Hardie's opinion,

From Amoeba to Man there has been a steady and more or less continuous progress. Some power has been at work seeking to make life perfect. . . . To say that . . . all the sweet and holy influence of Nature is the product of blind materialistic and unguided force is to me at least, unthinkable.[26]

More particularly, many found in Darwinism a justification for the strategy of British socialism — non revolutionary and relying upon the peaceful acquisition of state power by parliamentary means.

Ramsay MacDonald for example borrowed from Ritchie the organic analogy to argue for the need for the state.

The function of the nervous system is to co-ordinate the body to which it belongs . . . . Evidently the individualist cannot admit any such differentiated organ in society. But the socialist, on the other hand, sees its necessity. Some organ must enable the other organs and the mass of Society to communicate impressions and experiences to a receiving centre, must carry from that centre impulses leading to action, must originate on its own initiative organic movements calculated to bring some benefit or pleasure to the organism. This is the Socialist view of the political organ on its legislative and administrative sides.[27]

This view of the state led to an adulation of certain of its functions; in particular, its legislative role. Law was for MacDonald the ultimate realisation of the common purposes of society. It was therefore the highest expression of the moral instinct. Since the moral instinct was also the most important factor in the struggle for existence, in Macdonald's view, this made law the highest expression of evolutionary development. Basically MacDonald's view of evolution was an apologia for piecemeal parliamentary reform and for the idea that socialism can be enacted through the legislative functions of parliament.

This naturally separated his social Darwinism from that of the more radical socialist sects. According to a section of these, MacDonald's view of socialism neglected the importance of class struggle and revolutionary action. In their opinion this had no warrant in evolutionary biology.

It would be just as fair for a revolutionist to infer from this that Nature works only by revolutions as it ever was for an opportunist reformer to infer from this that Nature works only by evolution.[28]

The revolutionist could point, in fact, to

a collective form of struggle as well as the struggle of the individual. The various colonies of insects and animals such as ants, bees, wasps and beavers, discover to us a process that the human species could well imitate— I mean in a collective capacity— to rid themselves of a parasitic class that battens on their productive labour.[28]

When Mendelism became more widely known this version of socialism enjoyed, briefly, the discomfiture of the gradual evolutionists. Mendelism was interpreted by some biologists to mean that species evolved not by gradual accumulation of variations but by a qualitative leap in their character— in other words by mutation. This was received with some satisfaction by a number of socialists since it seemed to presage an analogous process in social evolution. According to the *Social Democrat*,

For a quarter of a century Socialists who had no comprehension of Science and scientists who had no comprehension of sociology, have kept us Revolutionists busy explaining that we were not unscientific and in conflict with the evolutionary theories of Darwin and Spencer because we believed a social cataclysm or revolution to be necessary . . . to break the shell of capitalism.[28]

In the opinion of the *Social Democrat*, Mendelism justified their opinion about the character of the evolutionary process.

In contrast, MacDonald replied that,

An attempt has been made . . . to give the class war a biological meaning. In industrial society, it is said, the struggle for life is carried on not so much between individuals as between classes, the bourgeoisie and the proletariat, the exploiter and the exploited. This does not correspond with the facts, for the more clearly economic lines are drawn between classes, the more intense becomes the struggle for life *within* these classes. What is in reality the most significant change in the struggle for life as seen in society is, that the individual struggle is no longer against nature but against a social organisation. In pre-civilised days man struggled with man and nature for subsistence which was scanty because nature was niggardly or unwooed by human toil; today man struggles with man and Society for subsistence which is scanty because the organisation of Society prevents the plenty which exists from finding its way into the possession of industrious men. The class struggle is not biological at all.[29]

Marxism failed as a science, argued MacDonald, precisely because of its lack of true acquaintance with Darwinism. The problem of Marxism was its basis in German Hegelianism, Hegel made transformation by the unity of opposites the means of evolutionary change in contrast to the Darwinian idea of a succession of evolutionary stages.

But if the socialists quarrelled among themselves over the existence of class struggle and its justification by reference to biology, they quarrelled with the anarchists about the state. Kropotkin's ideal was the stateless society which excluded precisely those institutions MacDonald defended.

In his book *Mutual Aid* (1902), Kropotkin took the argument back to Darwin's *Descent of Man*. Kropotkin argued that Darwin had shown that the greatest evolutionary advantage enjoyed by a species was 'sociability'. Hence the instinct of altruism gradually replaced that of self-interest. This was achieved without the diminution of natural forces — as Huxley suggested. On the contrary, it was nature itself which brought about sociability. Nor was the intervention of social institutions necessary to mitigate struggle and move evolution onto a higher path as Ritchie and Huxley believed. Far from this being the case, it was the intervention of social institutions which disrupted the harmony of nature.

There is a curious parallel between the conclusions which Spencer and Kropotkin drew from their very different descriptions of the natural world. They shared yet another characteristic — both wanted to restore the social world to a closer relationship with the natural one, as they saw it. In addition Spencer and Kropotkin believed that the means to do this was by removing those institutional organisations such as the state which stood between man and nature. The difference between them rests on the conception each had of nature. For Spencer, at least during the mood in which he wrote the articles for the *Contemporary Review*, the natural world was one of unrestrained competition and extinction of the unfit whilst, for Kropotkin, the natural world was co-operative and harmonious.

This mutual dislike of the state was recognised. It even led some anarchists to put forward the view that,

The State Socialists themselves are to constitute the raw material out of which is to be cultivated the national conscience. So that when Herbert Spencer described this sort of thing as the 'coming slavery' he wasn't so far wide of the mark'.[30]

For both liberals and anarchists the natural world was a self-regulating entity which required no special institution for its proper functioning.

Thus Darwinism served to support a wide range of political ideas from Left liberalism to socialism. For anarchists and liberals it showed nature and society required no intervention from without or from within by special institutions raised above the citizen. To Marx and Engels it revealed the bankruptcy of teleology in science in general. To a wide range of socialists it demonstrated the possibility of change in society and human nature and it acted as a challenge to religious authority. To others, it showed the necessity for gradual evolution presided over by the benevolence of the state and its functionaries and bolstered by a new ethic of citizenship. That this last version of socialism — discarding its social Darwinist elements — won out is another story for another kind of history. But it was the integration of socialism and liberalism which provided its intellectual character and its brief 'Darwinian' phase.

V THEORIES OF HEREDITY: LAMARCKIANISM

NINETEENTH-CENTURY social theorists believed in the importance of investigating the faculties of man. They hoped that by this process some clue could be found to the character of social evolution. Inevitably this brought up the question of heredity and the role it played in social life. Unfortunately there was no unanimity among biologists about the nature and cause of heredity. This situation lasted until the second decade of the twentieth century when the theory of Mendelism became generally accepted. Mendelism itself was not widely known until the early years of the twentieth century although it originated in the 1860s. Before then biologists had to make do with a number of competing theories of the character of heredity, all of which eventually proved to be, at the most, only of partial scientific value.[1] Biologists even moved uneasily between different theories. Darwin in 1859, on the publication of the *Origin*, took, on the whole, an anti-Lamarckian position. In the *Descent* in 1871 he had become much more favourable to Lamarckianism and this strengthened as he grew older.[2]

Lamarckianism was also called use-inheritance. It put forward the idea that changes in heredity occurred through the effort of the organism to adapt itself to changed conditions in its environment. In contrast to this Darwin in the *Origin* had argued that there existed a high degree of spontaneous variation in heredity regardless of environmental conditions. Darwin was not, however, able to prove how this occurred and this was one of the reasons why he was thrown back upon use-inheritance as an explanation of the origin of hereditary change.

This raised problems for his theory. As Spencer in the *Principles of Biology* in 1864 pointed out, if organic variation resulted from environmental pressure, then natural selection would be reduced to a subsidiary role in evolution. Spencer in his passages on natural selection in *Principles of Biology* pointed out how it would become merely a process of 'weeding' out changes which had originated by quite different principles of evolution. So therefore, whilst Darwin puzzled over the origin of variations and the influence upon them of the environment, he could never attribute them completely to it. This caused him to oscillate between Lamarckianism and the notion of 'spontaneous' variation and, as historians of his work have pointed out, sometimes placed him in an invidious and self-contradictory position.

For social theorists, however, Lamarckianism had a number of advantages. It was a means of explaining the origin and persistence of certain social customs and behaviour. An environment could be social as well as natural. In a particular social environment individuals might adapt their behaviour to the institutions of their society and gradually acquire an hereditary disposition for certain attitudes and behaviour. This explained why social behaviour 'fitted' society or, at least, did not diverge too radically from it. Moreover, if social behaviour followed the laws of natural selection then 'useful' social customs would survive just like 'useful' physiological characteristics.

This made use-inheritance a popular theory of heredity among mid-nineteenth century social thinkers. Darwin too used it to describe the origin of social customs. But in doing so he exhibited a great deal of caution. In the first place he did not consider many social customs to be useful. Some he considered completely idiosyncratic. How could these have become hereditary?

My chief source of doubt with respect to any such inheritance is that senseless customs, superstitions . . . ought on the same principle to be transmitted. Although this in itself is perhaps not less probable than that animals should acquire inherited tastes for certain types of food, or fear of certain foes, I have not met with any evidence in support of the transmission of superstitious customs or senseless habits.[3]

However, these doubts were not shared by others. Lamarckianism seemed to solve so many problems of social theory. It united biology with sociology. It explained how evolutionary change and social behaviour were linked. Moreover, it was a better mechanism for explaining this than natural selection. This had to await for a fortuitous variation to arise. The alternative theory of use-inheritance emphasised the direct influence of a social environment. This was, in Herbert Spencer's opinion, the best guarantee that behaviour and society could be rapidly brought into correspondence with one another.

If a nation is modified *en masse* by transmission of the effects produced on the natures of its members by those modes of daily activity which its institutions and circumstances involve; then we must infer that such institutions and circumstances mould its members far more rapidly and comprehensively than they can do if the sole cause of adaption to them is the more frequent survival of individuals who happen to have varied in favourable ways.[4]

Walter Bagehot adopted use-inheritance to explain the emergence of national character — a concept which had exercised a fascination over him from his earliest acquaintance with politics. For example, he regarded the French Revolution of 1848 and the subsequent events of 1848–51, some of which he witnessed personally, as exemplifying a typically French attitude to political institutions. This was characterised by,

a certain mobility that is, as it has been defined, a certain excessive sensibility to *present* impressions, which is sometimes levity; for it issues in a postponement of seemingly fixed principles as to a momentary temptation or a transient whim; 'impatience' as leading to an exaggerated sense of existing evils.[5]

At that time he confessed himself at a loss to find the reasons for its emergence.

The formation of *this* character is one of the most secret of marvelous mysteries. Why nations have the character we see them to have is, generally speaking, as little explicable to our shallow perspicacity as why individuals, our friends or our enemies, for good or for evil, have the character they have.[5]

According to Richard Holt Hutton, Bagehot had a 'keen sympathy with natural science and devoured all Mr Darwin's and Mr Wallace's books and many of a much more technical kind'.[6] Certainly he read Maudsley and Carpenter both of whom incorporated use-inheritance into their psychological theories by the 1860s. Therefore when Bagehot came to write the articles in the 1860s which were eventually collected into *Physics and Politics* he had found a scientific underpinning for his theories of character. In *Physics and Politics* Bagehot quoted Maudsley's opinions about,

The way in which an acquired faculty of the parent animal is sometimes distinctly transmitted to the progeny as a heritage, instinct or innate endowment. . . . Power that has been laboriously acquired and stored up as statical in one generation, manifestly in such case becomes the inborn faculty of the next.[7]

Bagehot connected this process with the formation of national character in a number of ways. Firstly, the experience of living in a particular social environment formed habits and attitudes which became hereditary. However, Bagehot also allowed a place in the formation of national character for the outstanding individual.

In Spencer's sociology the great man was merely an obstacle to understanding the processes of social evolution. Bagehot, on the contrary, regarded the outstanding individual as one who, by dominating the manners and customs of his society— even in matters of fashion— succeeded in impressing his or her character on national character at large. The mechanism by which this took place was imitation. The organism's reaction to its environment paralleled the individual's imitation of great men and women. In both cases a hereditary faculty might result. Darwin himself used this notion. For in Darwin's theory of sexual selection, the particular physical features of races arose from a conception by different societies of an ideal of beauty and an attempt to emulate this ideal among its members. This process arose in Darwin's case by natural selection rather than use-inheritance. However, it is possible to see how social theorists could argue for a similar process in the origin of a society's manners and customs. Thus Bagehot argued that the sober appreciation of the process of parliamentary government among Englishmen and the expression of this in the policies of certain great statesmen provided an ideal which moulded the character of the English nation. The admiration for rationality among the French gave rise to their own tendency to draw up formal constitutional rules. Moreover, the people's respect for the monarchy was the means by which its values might become the heredity of the British people in every sense of that word. A tendency to reverence could, by example, become an indelible inheritance in the English working class.

In this way use-inheritance certainly solved the problem of social order. It also seemed to hold out the prospect of evolutionary progress. For hereditary improvements could accumulate over time. Bagehot felt that use-inheritance was, 'a continuous force which binds age to age which enables each to begin with some improvement on the last . . . .'[7] Darwin also hoped that use-inheritance might be the basis of social progress.

Looking to future generations, there is no cause to fear that the social instincts will grow weaker, and we may expect that virtuous habits will grow stronger, *becoming perhaps fixed by inheritance.*[8]

This hope in the improving power of use-inheritance was largely based upon the belief that the benefits of education could actually impress themselves on man's hereditary character and be transmitted by inheritance to succeeding generations. As Darwin wrote to a correspondent,

If it could be proved that education acts not only on the individual but, by transmission, on the race, this would be a great encouragement to all working on this all-important subject.[9]

W. B. Carpenter also believed that use-inheritance was 'our surest evidence of the influence of the culture of one generation upon the *thinking power* of the next'.[10]

This meliorism proved an enduring characteristic of Lamarckianism. It has led some to regard it as the theory of heredity most appropriate to schemes of social improvement. However, this relationship was not as simple as might at first appear. In an article in *Mind* in 1876 on comparative psychology, Spencer showed how use-inheritance could produce a biological determinism as strong as other theories of inheritance. Certainly institutions moulded character but the effects of this moulding might be to trap an individual's heredity in a set of permanent roles or faculties from which they could escape only — if ever — after a long historical process. Thus Spencer's comparative psychology was a highly stereotyped view of the relations between women and men and also between the races. By suggesting that institutions produced hereditary changes in character and that character was not a product of external interventions by social institutions in the lifetime of an individual, use-inheritance might give less freedom to the individual to change his or her pattern of behaviour.

This led some social theorists, contemporaries of Darwin, Spencer and Bagehot, to reject it. For example, Stephen and Pollock both considered it inappropriate to their social theories. Pollock in 1876 argued that, in connection with the social and moral ethic,

... it is at least a rational question whether the specific working of those instincts . . . passed on by tradition to the succeeding ones (generations) are not, to some extent, reinforced by physical inheritance.[11]

On the whole, however, Stephen, Pollock, Hobhouse and Ritchie remained immune to use-inheritance. They tried to incorporate a theory of moral voluntarism into evolution. Lamarckianism could be combined with this but it also confused it. How, for example, could the idea that change was provoked by external conditions be reconciled with the notion of the individual as the source of moral and intellectual innovation? Similarly, at what point did adaption by the intelligence and will of the organic form become the mechanical pattern of hereditary behaviour in its descendants?

There was therefore a degree of contradiction between the meliorism of use-inheritance and a potential determinism in it. Consequently Stephen rejected it even whilst it was still a fashionable talking point in social theory. Instead he put the whole burden of the reproduction of social behaviour onto education, primarily within the family, and upon the inner sense of responsibility that education in moral values was supposed to provide.

But not only was Lamarckianism unconvincing to many as an explanation of social behaviour; its value considerably decreased under the impact of the criticisms of the German naturalist August Weismann in the 1880s. Weismann's work on heredity became available in English in 1882 and was discussed at the British Association of 1887.[12] Weismann had subjected use-inheritance to a series of tests and come up with the conclusion that the notion of inheritance of acquired characteristics was not founded upon experimental proof. In place of it he developed the theory of the 'germ plasm' of an hereditary material passed on from generation to generation and largely unaffected by the environment. By the 1900s Lamarckians were a minority within biology.

Weismannism in its turn became integrated within social theory. Francis Galton regarded the dismissal of use-inheritance by Weismann as support for his own views on the all importance of heredity. But it was also clear that the ameliorism rather than the determinism of use-inheritance made it the object of attack in the late nineteenth and early twentieth century. Many defences of social reform were, of course, made on other grounds than the effect such reform would have on the hereditary constitution of the race. Nonetheless there was a certain undercurrent of dissatisfaction among some social reformers at the time about the lack of a direct biological argument for social welfare measures. L. T. Hobhouse, for example, in the confusion which arose in the early twentieth century among Weismannites due to the introduction of Mendel's principles of heredity, hoped that out of the revision of studies of inheritance in

the plenitude of our ignorance as to variation . . . there are not wanting indications that the environment had indirect and subtle effects which had yet to be measured. We shall have to know more of the response of racial types to new surroundings and of the mechanism by which this response is effected before we can be sure that, not indeed by the direct transmission of acquired characters but by some far more subtle series of spontaneous responses to new stimuli, the race does not adapt itself, as a race, to changed conditions whether for good or ill.[13]

There were other attempts to 'humanise' the theory of Weismannism. For example, in *Social Evolution* (1894), Benjamin Kidd put forward an ingenious argument derived from Weismann to justify an altruistic basis for human social behaviour and the ultimate value of measures of social reform. It was based on the hypothesis that the more complex a biological organism, the shorter was its duration of life. Consequently, these shorter lived organisms were able to pass on the accumulated hereditary improvements which occurred during their span of life to their descendants at a faster rate than the less complex and longer lived organisms. Hence the rate of genetic improvement in these species occurred at an accelerated rate compared to other species.

Claiming the authority of Weismann, Kidd transposed this argument into his theory of social evolution. The rate of social improvement and efficiency is increased in that social organism in which the interests of the existing members of society are subordinated to those of its future members. This will manifest itself in their acceptance of the struggle for existence and the penalties which this imposes. At the same time it will lead to the concentration of effort upon measures of social improvement of which the major beneficiary will be future generations. This will involve abandoning self-interested calculation as the basis of social action. But although Kidd felt he could reconcile his theory of social evolution and Weismann's theory, Spencer was in a more difficult position.

Spencer's contemporaries were well aware of the effect Weismann's attacks on Lamarckianism would have on his theories. Wallace wrote to a colleague that,

Thistleton-Dyer tells me that H. Spencer is dreadfully disturbed on the question. He fears that acquired characteristics may not be inherited, in which case the foundation of his whole philosophy is undermined![14]

Spencer's philosophy was based on the notion of persistence of force which was derived from physics rather than biology. This stated that an object redistributed its parts under the impulsion of an external force bearing on it. Lamarckianism represented this process in the natural world. Its major features were easily assimilated to it. The object was the natural organism; the external impulsion was changes in the environment; the process of readjustment was the adaption of the organic form to these changes. The ultimate result of this was organic evolution. The whole edifice of Spencer's systemised philosophy rested upon the correlation between different sections of it. What was at stake for Spencer was not just a particular theory of character but part of his claim to have produced his correlated system.

Spencer's reply to Weismann was *Factors of Organic Evolution* (1887) and a long controversy with Weismann himself in the pages of the *Contemporary Review* in the 1890s. In these Spencer conceded a greater equality to natural selection than he had in his earlier work. Use-inheritance was, however, retained even though Spencer now admitted that its effect was only on a 'minor part of the facts'.[15] Spencer in these works also brought up what was to be a constant theme of Lamarckians — the fact that Darwin himself had never completely abandoned use-inheritance. He complained, in fact, that,

Nowadays most naturalists are more Darwinian than Mr Darwin himself. I do not mean that their beliefs in organic evolution are more decided . . . . But I mean that the particular factor which he first recognised as having played so immense a part in organic evolution has come to be regarded by his followers as the sole factor though it was not so regarded by him.[15]

Prince Kropotkin, like Spencer, continued to defend use-inheritance. His political philosophy depended upon the notion of a rapid transformation of human nature by the reform of social institutions. This included the removal of those which corrupted the peaceful, co-operative and industrious individual upon whom the stateless society depended for its viability. In a series of articles in 1910 he adduced a number of elaborate scientific proofs in defence of Lamarck. Like Spencer, he attempted to amalgamate natural selection and use-inheritance.

It is self-evident that those biologists who recognise the dominant influence of a direct action of environment do not necessarily deny the intervention of Natural Selection. On the contrary, they fully recognise its usefulness as an auxiliary. They only limit its powers. It ceases to be a selection of *haphazard variations — necessarily indifferent in most cases in their incipient stages* — but becomes a physiological selection of those individuals, societies, and groups which are best capable of meeting the new requirements by new adaptations of their tissues, organs, and habits . . . . It is also, in the animal world, a selection of those who best exercise their collective intelligence for the diminution of inner competition and inner war, and for the rearing of their offspring by combined effort. And finally, as it does not depend upon an acute struggle between all the individuals of a group, it does not require for its full action those exceptionally bad seasons, droughts, and times of calamities which Darwin considered as especially favourable for Natural Selection.

This again brought Kropotkin close to Spencer. Certain social theorists like Stephen saw the adaption of character to society as the effect of intervention in behaviour from external forces. Stephen saw this force as the family. Ritchie argued that the intervention of the state helped bring into existence certain potentialities in human behaviour. In these cases the continuation of these social institutions was necessary because they produced no permanent hereditary effect. But both Kropotkin and Spencer hoped that character itself — by the internal mechanisms of heredity— might bear the accumulated culture of generations and render unnecessary some at least of these institutions.

II

In spite of Weismann's criticisms, use-inheritance survived. It did so partly because no theory of heredity appeared to answer all the problems of inheritance. Lamarckians were generally always able to point to inconsistencies and obscurities in other theories of heredity. Since most presupposed that theories must, by their nature, illuminate all adjacent areas, this was considered sufficient reason by many to cling to use-inheritance as an 'explanation' of the areas which still remained shrouded in mystery.

But there were other reasons for its persistence. Some social theorists wanted heredity and evolution to be combined with psychical or psychological causes — in particular with notions of consciousness, intelligence and individual choice. These concepts were the stock-in-trade of psychological analysis and many psychologists were insistent that the theory of natural selection itself should find room for these notions. In contrast to the notion of random and fortuitous variation implied by natural selection, Lamarckianism suggested that variation originated in an individual's reaction to his environment— in some sort of striving to adapt. It put psychological and behavioural processes at the centre of evolution. It could also be combined with a theory of the role of intelligence. If intel-

ligence played a part in this striving for adaption then the more
intelligent a creature the more successful would its adaption
be. G. J. Romanes, to whom Darwin left his psychological
papers, produced a modification of the theory of natural selection which included this proposition. Romanes had argued
that the application of Lamarckianism meant that,

natural selection is not left to wait, as it were, for the required variation to
arise fortuitously; but is from the first furnished by the intelligence of the
animal with the particular variations which are needed.[17]

J. M. Baldwin, an American psychologist, followed Romanes
in revising the theory of evolution to include a variation of
Lamarckianism. He attempted to persuade the more orthodox
Darwinians that they too should do this. Baldwin addressed
his remarks firstly to Wallace. This had a peculiar piquancy
since Wallace had originally, in 1864, suggested that consciousness played a greater part in human evolutionary change
than other factors and Baldwin was quick to point this out.
Baldwin, in defending this theory to Wallace, pointed out that
it gave,

consciousness a place even in the lower forms, inasmuch as it considers
consciousness as a great 'accommodation' (or adaption) agent. . . . How do
you otherwise bring consciousness into the workings of the process of
evolution. I find mental endowment as critical as supplementing all sorts of
physical characters.[18]

But Wallace, although he conceded that he had brought up the
question of consciousness in evolution, was not prepared to
revise the theory of natural selection

Your account of Organic Selection, as originated by yourself and Lloyd
Morgan is very clear and I have no doubt it is occasionally a real factor in
evolution. But I do not think that it is an important or even an essential one
. . . all the arguments of H. Spencer and others as to the impossibility of
co-incident variations of the right kind occurring *when required* seem to be
purely verbal objections not warranted by the facts of nature.[18]

Theorists like Baldwin thought the chances of an appropriate
variation occurring spontaneously was small but that this
problem would be overcome if the individual could con-

sciously strive to meet the demands of its environment. These theories were a modification of use-inheritance away from emphasis upon the moulding effects of the environment to the role of the organism, in many cases, initiating change. Those who see Lamarckianism purely as environmental determinism are, in fact, seeing only one aspect. In the hands of the psychologists this aspect was increasingly subordinated to the others.

Darwin's contemporaries had used Lamarckianism to biologise associationism. They suggested that complexity arose out of the process of action and reaction to the environment experienced by an organism. But psychologists like Piaget argued, in contrast, that a pre-determined complexity existed in the organism which allowed it to manipulate its environment rather than the obverse. What brought this complexity into existence was not just the experience of environmental stimulus but maturation — a process internal to individual organic development. Piaget was willing to concede that this complexity may have arisen as the result of a long evolution in which simple constituents of behaviour gradually developed into more complex ones. But this was only of historical interest for his psychology.

Piaget's case shows how a form of use-inheritance could be combined with a radical anti-environmentalism. The environment which determines heredity is the organism's *internal* environment into which, in some way, external experience is fed by the choices and manipulation of the individual.

There was, in addition, a strong philosophical and even religious resistance to the notion of fortuitous variations — in the sense of variations not directed by some purpose, either that of God or the individual. Some thinkers were unable to accept the notion that chance or fortuitous undirected variation was the material out of which evolution progressed. This seemed to them to replace a universe of order with chaos. Use-inheritance was able to express this element of moral discontent with the theory of natural selection and its implications. This surfaced very quickly after the publication of the *Origin*. St George Mivart, the naturalist and Catholic apologist, felt that the *Origin* struck a blow at the basis of religious belief not simply because it supported evolution but because of the character of this evolution. The classic theory of natural selection (before Darwin made his concessions to Lamarckianism) rested on the notion of fortuitous variation occurring at random. In Mivart's opinion this denied that man and nature had any moral or religious purpose.

there is fully as real a distinction between the production of new specific manifestations entirely *ab externo* and by the production of the same through an innate force and tendency.[19]

There were, however, theories which retained a place for 'an intelligent and self-conscious will'[19] in evolution and although in Mivart's case use-inheritance did not necessarily satisfy completely, Lamarckianism was adapted for this purpose by other writers.

The tradition which saw the random variation robbing life and history of purpose stretches from Samuel Butler to Bernard Shaw. Shaw, in the Preface to *Back to Methuselah* in 1921 expressed this revulsion perfectly.

What hope is there then of human improvement? According to the Neo-Darwinists, to the Mechanists, no hope whatever, because improvement can come only through some senseless accident which must, on the statistical average of accidents, be presently wiped out by some other equally senseless accident.[20]

By the 1900s this discontent coalesced with the crisis in Darwinian science which accompanied the introduction of Mendelism into Britain. In the early 1900s Mendelism was introduced into Britain mainly through the naturalist William Bateson who set out the principles in a number of books and articles. Bateson was, however, anti-Darwinian. He believed that Mendel's laws of heredity showed that species were formed — in contrast to the emphasis laid by Darwin on small, slight variations in the character of organisms — by larger jumps in heredity or mutations. This view was later proved incorrect and the theory of natural selection was shown to be compatible with Mendelism which extended the possibilities of natural selection as a form of explanation of evolutionary change. Nonetheless, at the time, Bateson and other naturalists were genuinely puzzled as to how the two — natural selection and Mendelism — could be integrated and this led many to regard natural selection as a superseded theory.[21]

This intervention produced a strong reaction from two sources. Firstly, the orthodox Darwinians were outraged at Bateson's interpretation of Mendelism. Wallace in particular was moved to pen protests at Bateson's views. In support of his attack, he could show that Bateson in the 1890s, before Mendel's work had become known, had based his theory of evolution on mutation theory. Wallace had written to Poulton in 1894 complaining about Bateson that,

> Neither he nor Galton appears to have any adequate conception of what Natural Selection is, or how impossible it is to escape from it. They seem to think that, given a stable variation, Natural Selection must hide its diminished head.[22]

This was a fair point. The intricacies of the theory of natural selection were lost on many of its critics. Most of the problems which Mendelism seemed to pose for Darwinian evolution were solvable but Bateson saw Mendelism as an alternative to natural selection and as adequate in itself to account for the formation of species. In a sense Bateson could be accused of adapting Mendelism to suit preconceptions he already had in the 1890s of the origin of species.

Wallace's anger consequently increased and in 1905 he wrote to Sir J. Hooker that,

> F. Darwin lent me Prof. Hubrecht's review from the *Popular Science Monthly* in which he claims that de Vries has proved that new species have always been produced from 'mutations', never through normal variability, and that Darwin latterly agreed with him. This is to me amazing![22]

But the orthodox Darwinians were also highly suspicious of the quasi-religious element in the welcome given by some philosophers and biologists to Mendelism. The problems of Darwinism were rapidly widened to include a general attack on Darwinian 'mechanism'. E. B. Poulton, for example, voiced his suspicions when in a largely scientific defence of Darwin in 1907 he wrote,

> Uninstructed statements — commonly encountered just now in the press —

inform the world that Natural Selection is entirely dispensed with by modern writers on Mutation and Mendelism. This is of course an error. Mutation without Selection may be left to those who desire to revive Special Creation under another name.[23]

In 1911 Henri Bergson gave several lectures in England at Oxford, Birmingham and London. In the same year *Creative Evolution* (1907) was translated into English and published. This had considerable effect upon the debate about evolution and heredity. Basically the force behind evolutionary change was, in Bergson's opinion, a creative will which he called the 'élan vital'. Bergson put forward the view that it was this creative will and striving in the organic form which caused heredity to reconstitute itself in new forms. This evolution was self or inner directed only to a limited extent. The 'élan vital' took on the character of some universal life force which expressed itself in various departments of life not just evolutionary change. Also, according to this view, there was no pre-ordained 'end' to evolution. The creativity of evolution was such that no future prediction of the character of evolution could be made.

Bergson's relation to the existing controversy in Britain had various aspects. Firstly, he agreed with Samuel Butler on the inadequacy of natural selection to explain evolution. This view Bergson repeated even in the 1930s. Against the evidence of the work done by R. A. Fisher in the 1920s in integrating Mendelism and Darwinism, Bergson claimed that 'cette insuf-fisance du darwinisme originel est depuis longtemps recon-nue par la presque totalité des biologistes et des philosophes'.[24]

Nonetheless Bergson was not a Lamarckian. He gave Lamarckianism only partial approval. In so far as Lamarckian-ism appeared to introduce a psychical element into evolution, Bergson approved. According to Bergson, 'Neo-Lamarckism is therefore, of all the later forms of evolutionism, the only one capable of admitting an internal and psychological principle of development although it is not bound to do so'.[25] Nonetheless use-inheritance was still a 'mechanical exercise of certain organs mechanically elicited by pressure of external circums-tances'. The virtue of the 'élan vital' was that it was 'even more psychological than any neo-Lamarckian supposes'.[25] This was because it excluded environmental influence from the process of evolutionary adaption, relying instead on the life force as the agent of change.

The conjecture of Bergsonism and Mendelism resulted in a peculiar intertwining of philosophical and scientific issues. It was easy to suppose that the theory of evolution by mutation and Bergson's 'élan vital' complemented each other. The impetus behind evolution by mutation was interpreted as emanating from the life force. Graham Wallas, the political scientist, wrote to Shaw several years after the controversy began and shortly after Shaw's Preface to *Back to Methuselah* appeared, putting forward a summary of views of this sort.

In a few weeks time I shall be sending you a book of mine, in which I argue for the necessity of conscious purpose as against half-conscious drift in human affairs. Before saying what I want to say about your preface may I put my own position on the biological points. Twenty years ago I was a pure Darwinian, believing that natural selection turned small chance variations into new species. Now I learn from the biologists that mere 'variations' . . . don't count. What matters is, they say, 'mutations', big, sudden hereditary changes. Bateson and the micro-sophists seem to be learning a good deal about the manner of mutations (Mendelism etc) but no-one has any general explanation of the 'cause' of them. They may well be the result of a general drive in things which one can call 'vitalism' or 'élan vital' — so far I can take your position as a working metaphysical hypothesis and life, 'élan vital', 'will' and 'desire' may all be the same thing.[26]

Although a variant of the 'élan vital' was acceptable to Wallas, Lamarckianism was not. However, in the minds of many commentators it was not so easy to separate the idea of consciousness and intelligence in evolution from use-inheritance. Bateson referred to this when he wrote in 1915,

It is true that in the last decade some have again revived the view brilliantly expounded by Samuel Butler (*Life and Habit* 1878) and also by Hering, that living things may, through their generations, have a continuous accumulation of 'unconscious memory'. Just as learning to read or play a musical instrument requires close attention and extreme effort in the early stages — though afterwards the acts may be performed without conscious attention at all — so it is argued may even the ordinary reflex actions . . . have been acquired as a summation of effort originally conscious. . . . But since, as we have said, there is no good reason to suppose that even the simplest experiences of the parent are at all transmitted to a succeeding generation, the suggestion of continuous memory as applicable to education can only be defended on grounds which to the biologist are mystical and unconvincing.[27]

There were two sources of opposition to the belief in heredity by random variation. Both were ultimately political and social objections. One thought that random variation excluded planned progressive improvement in human heredity and felt that by losing this theory of heredity, they had lost a substantial argument in favour of social change. The other strand of opposition was more concerned with religious, mystical and philosophical questions about purpose and direction. This last strand did not necessarily hold out the hope of social progress. It could, in fact, regard life as essentially unchanging and unimprovable. It did, however, insist upon the *directed* character of evolution — often by man but also often by divine intelligence or mystical force. Kropotkin, who held the former view, was well aware of this second kind of appropriation of Lamarck and in 1910 swiftly stepped in against these metaphysical interpretations.

As to the exaggerated 'interference of the animal's will in the formation of new organs', of which metaphysically inclined writers have lately tried to make so much, Lamarck distinctly said . . . that variation in plants is fully due to change in food, in absorption and transpiration, and in the quantities of heat, light, air, and moisture received. 'Plants have no will.' And as to animals he insisted, repeating that it is only in insects and the classes superior to them that 'sensation and effort', originated from a need, can be effective in producing new *habits* which will contribute to modify structure — it being the function which creates the organ, not the reverse.[28]

Nonetheless many writers saw Mendel, Bergson and Lamarck in a continuum with the emphasis on the death this would spell to Darwinian mechanism. C. W. Saleeby, for example, argued that,

Darwin himself always believed, with his predecessor Lamarck, whom the French justly regard as the real pioneer of organic evolution, that the willing adaption of individuals to their conditions of life was reflected in their offspring, so that life became more apt and more secure in its manifestations from generation to generation. Here there is recognition of a positive factor which is not mechanical but psychical. Everyone who still echoes the dead materialism of the nineteenth century will be aghast, no doubt, but we must go forward — in the illustrious company of such leaders as Bergson in Paris, Driesch in Heidelberg and McDougall in Oxford.[29]

To many, the social meaning of the Bergson–Mendel couplet

was the recurrence of the old struggle between Mivart and Darwin. In addition, although Shaw gave the life force a 'socialist' interpretation — in Shaw's view the life force was the creative current which destroyed the old institutions and created the new — many believed it was a reaffirmation of the unchanging and unalterable character of social life. It emphasised the links with past, the persistence of certain basic forces in human nature and society and, to the secularist, it was a new and aggressive defence of religion. This was confirmed by the title of the lectures Bergson gave in London on 'The Nature of the Soul' and by the numerous celebrations in the periodical press and newspapers of the demise of Darwinian 'mechanism'. The secularist Hugh S. Elliott referred in 1913 to

journalists, sciolists, demagogues and popular lecturers at large who have been steeple chasing over the country declaring that mechanistic and materialistic biology was dead.[30]

The orthodox Darwinian scientists were moved to protest at Bergson's attack on Darwin. E. B. Poulton attempted to refute Bergson's idea that the existence of instinct proved an intuitive understanding existed in the organism. In what was very much a repetition of the struggle Darwin had had years earlier, E. B. Poulton tried to show how intricate patterns of behaviour in animals could be explained by natural selection rather than by Bergson's 'hypothesis of a creative "internal developmental force" '.[30] The zoologist E. Ray Lankester, too, felt that the issues of spirit and Mendelism had been unfairly entwined. He added a preface to the book by the secularist Hugh S. Elliott attacking Bergson's doctrines.[31]

The period inaugurated by Bergson's emergence in England served to sharpen all the issues which had accompanied the original publication of the *Origin*. By that time, however, the battle lines were drawn not between revealed religion and science, although this played a part, but between philosophical tendencies. On the one hand was the insistence that a psychical element of whatever kind — Bergsonist or Lamarckian — existed in evolution. On the other was the insistence that it did not. McDougall, for example, claimed that,

Those who speak of the mechanical working of natural selection forget the 'struggle for existence' which is essentially a psychical struggle in that it presupposes the 'will to live'.[32]

Whereas Hugh S. Elliott, who attacked McDougall's *Body and Mind* as a sophisticated defence of animism, argued that,

Natural Selection not only does not require a psychical factor but definitely excludes the possibility of it, for the insect does not struggle to look like a dead leaf, nor does it thereby display any 'will to live'. The theory of Natural Selection is that it is wholly impassive.[32]

The other difference in this battle was that, apparently, science had intervened on the side of spirit. Or at least this was McDougall's belief. He considered that science had outgrown its materialist tendency. So, too, did C. W. Saleeby.

Seven years ago M. Bergson's 'Creative Evolution', the greatest book of our century, was published. When we return to the Darwinian theory, with our new genetic knowledge of the manner in which variations arise, and our new perception of the fact that their origin or creation is the crux of evolution, we are disconcerted to discover how little indeed it is that Darwinism really accounted for. Professing to replace the old doctrine of 'special creation' . . . . Darwinism offered us nothing but destruction. 'Natural Selection' is not a constructive process at all. It is not positive but negative — it is natural rejection not natural selection.[33]

This confidence, however, was short-lived. In the 1920s R. A. Fisher and others demonstrated how the new genetics and natural selection could be reconciled. This tended to push McDougall further towards a Lamarckian solution to evolution. In 1928 he advocated that,

we should go boldly back to Lamarck and assume with him that the essential factor to be investigating is the effort, the more or less intelligent striving of the organism to adapt itself to new conditions.[34]

In the middle of the twentieth century there were peculiar resonances of the debate. Shaw lived to see Mendelism and Darwinism integrated in the theories of R. A. Fisher and J. B. S. Haldane. However, in 1948 he interpreted the emergence of Lysenkoism in the Soviet Union as a long-awaited confirmation of his side in the scientific controversy of forty years previously. In Lysenko Shaw saw a reaffirmation of the role of mind in evolution. He considered that Lysenko had been anticipated eighty years ago by Samuel Butler who realised 'the enormity of the fatalism inherent in Darwinism'.[35] After Butler came Shaw and then Bergson, in that order. Moreover,

After Bergson, Weismannism lost its stranglehold on the scientific world. Scott Haldane (father of J.B.S.), Needham and in Russia Michurin and Lysenko broke away from Fatalism, not polemically, but by simply ignoring it.[35]

Shaw put one aspect of the case perfectly. Although careless in lumping together thinkers as diverse as Butler, Bergson and Lysenko, there was a degree of continuity which drew all three together. This arose from their challenge to theories of inheritance which expelled mind or environment from direct influence on evolution. This produced an undercurrent of opposition to the development of studies of heredity from the mid-nineteenth to the mid-twentieth century. It was one which showed a surprising tenacity. Even when it seemed that the advances in genetical theory had dealt it irrevocable harm, Lamarckianism had a tendency to reappear in a new form and in a new context.

This persistence was due to several factors. On the one hand a search for evolution which could be reconciled with religious or mystical interpretations; on the other the desire for evidence of the operation of human intelligence in evolution. Also involved were a section of psychologists who saw use-inheritance as enshrining the concepts of their discipline more satisfactorily than any other alternative social theory.

Finally, one other section of intellectual life — the theorists of social reform — provided another area from which defences of use-inheritance came. To these the randomness of variation seemed a very shaky basis on which to build a theory of progressive human improvement brought about by welfare and reform. It was possible, of course, to remove the argument about social reform from the area of biology altogether. However in the early twentieth century, this tended to be an unusual position. In part this was due to the influence of the eugenics movement who attempted to put the question of social reform and theories of inheritance firmly together. For social reformers to ignore eugenics and its claims about heredity in the early twentieth century would have been to opt out completely from one of the major controversies about social welfare.

VI THEORIES OF HEREDITY: THE EUGENICS MOVEMENT

THE inaugural address to the Association for Social Science which met in 1860 described the objects of the Society as,

to guide the public mind to the best practical means of promoting the amendment of the Law, the Advancement of Education, the Prevention and Repression of Crime, the Reformation of Criminals, the adoption of Sanitary Regulations and the diffusion of sound principles on all questions of Social Economy.[1]

The work of the eugenics movement exemplified this approach to social theory. Other social Darwinisms suggested a particular approach to politics or the importance of education and intellectual freedom. Eugenics claimed above all to be a guide to the actual administration of society and to provide the categories and principles upon which social legislation should be based. The consequence of this was that the history of social reform in the early twentieth century — when eugenics was at the peak of its influence — was closely bound up with the progress of the eugenics movement.

Francis Galton founded eugenics with the aim of raising the physical and mental level of the race. This idea was based on two propositions: first, that desirable physical and mental qualities were unevenly distributed throughout the population and, second, that those who had the desirable qualities could be identified and encouraged to multiply faster than the others. This, he argued, could be accomplished partly by various forms of state intervention and partly by placing a high social value on the fertility of the better stocks in society.

Galton claimed that eugenics was practical Darwinism. His intention was, he claimed, to 'see what the theory of heredity, of variations and the principle of natural selection mean when applied to Man'.[2] His first discussions of eugenics took place in the 1860s after the publication of the *Origin*. This event certainly increased in his mind the significance of differential fertility and heredity. His relation to Darwinism, is however, much more complex than the picture which both he and his followers drew.

According to the theory of natural selection there was only one criterion of the 'fittest'. These were the individuals who left the most progeny. Galton found it impossible to reconcile this with the fact that in the mid-nineteenth century the poorer classes appeared to be the most fertile. The struggle for existence, he wrote,

seems to me to spoil and not to improve our breed. . . . On the contrary, it is the classes of a coarser organisation who seem to be on the whole, the most favoured under this principle of selection and who survive to become the parents of the next.[3]

This idea Galton put forward in a series of letters to Darwin. They made Darwin distinctly unhappy. In the late sixties he was gathering material for the *Descent of Man* in which he hoped to prove both that natural selection could be applied to man and that it was still in operation in human evolution. He saw Galton's theories as largely another argument which could be adduced against this idea. Therefore, although he gave cautious support to Galton's view of the importance of intellect in human evolution, his references to the theory in the *Descent* were largely replies to Galton's view that natural selection no longer operated in human society. Therefore, according to Darwin,

It has been urged by several writers that as high intellectual powers are advantageous to a nation, the old Greeks, who stood some grades higher in intellect than any race that has ever existed, ought to have risen, if the power of natural selection were real, still higher in the scale, increased in number and stocked the whole of Europe.[4]

This was based in Darwin's view, on the erroneous assumption that,

there is some innate tendency towards continued development in mind and body. But development of all kinds depends upon many concurrent favourable circumstances. Natural selection acts only in a tentative manner. Individuals and races may have acquired certain indisputable advantages and yet have perished from failing in other characters.[4]

Darwin did not deny the importance of intellect in evolution though it was also his opinion that, 'excepting fools, men did not differ much in intellect, only in zeal and hard work; and I still think (this) is an eminently important difference'.[5] But he objected to the belief that if natural selection was not operating according to Galton's criteria of eugenic worth this must imply that it was not operating at all.

Galton had, in fact, substituted a highly subjective criterion of eugenic worth. Eugenics was a combination of the language of natural selection with highly partial and contentious social judgments on the relative worth of different sections of the population. Most of Galton's career was concerned with establishing apparently objective measures of worth. This ranged from the statistical distribution of intelligence to anthropometry — a series of tests, both physical and mental, to measure differences in faculty. The first attempt at measuring the statistical distribution of intelligence was made in *Hereditary Genius* in 1869. For this he used the obituary columns of *The Times*. Eminence, he believed, could be measured by the opinion of men held by their peers, but even here he made a number of reservations. He refused, for example, to include politicians for,

many men who have succeeded as statesmen, would have been nobodies had they been born into a lower rank of life.

Similarly, civil servants were excluded on the grounds that,

I do not, however, take much note of official rank. People who have left very great names behind them have mostly done so through non-professional labours. I certainly should not include mere officials except of the highest rank and in open professions among my select list of eminent men.[6]

These opinions reflect strongly the social prejudice of the nineteenth-century middle class against the dominance in politics of the aristocracy and against an unreformed civil service. Later in the century, a section of the eugenic movement attempted to rescue the aristocracy from the opprobrium of being a dysgenic class. This reflected a change in social attitude rather than a refinement in the techniques of measuring objective worth.

However volatile certain of the social judgments of eugenics were, there was a consistent element. This was the conviction that a high birth rate among the lower classes was a threat to evolutionary progress. This high birth rate far from impressing the middle classes with the evolutionary fitness of the poor tended to be seen as further evidence of their inferiority. According to one commentator,

The careless, squalid, unaspiring Irishman, fed on potatoes living in a pig-stye, doting on a superstition, multiplies like rabbits or ephemera; the frugal, forseeing self-respecting Scot passes his best years in struggle and celibacy.

This contempt was, in the early stages of eugenics also directed at the aristocratic class. W. R. Greg considered that,

Not only does civilisation, as it exists among us, enable rank and wealth, however diseased, enfeebled or unintelligent, to become the continuators of the species in preference to larger brains, stronger frames and sounder constitutions: but that very rank and wealth, thus inherited without effort and in absolute security, tend to produce enervated and unintelligent offspring. To be born in the purple is not the right introduction to healthy energy.[7]

But Greg also ventured the opinion that the chief threat to society lay in the fact that 'Malthus's "prudential check" rarely operates upon the lower classes, "the poorer they are usually the faster do they multiply".'[7]

When in 1904 the Committee on Physical Deterioration asked a witness whether, 'the better stocks, as represented in the upper and middle classes, are not reproducing themselves at the same rate as formerly?',[8] they were eliciting an answer which had already been formulated for many years. Galton's work was based on the opinion that this was the case and it was also the opinion of many commentators like Greg. Of this view a number of things should be said. In the first place it was bound up with the importance of Malthus in nineteenth-century political economy. Malthus had made population laws a political doctrine, part of the liberal, anti-socialist catechism, and even those most favourable to working class efforts to improve their conditions of life were influenced by his views. J. S. Mill, for example, confessed that, early in his life,

> the notion that it was possible to go further than this in removing the injustice . . . involved in the fact that some are born to riches and the vast majority to poverty, I then reckoned chimerical, and only hoped that by universal education, *leading to voluntary restraint on population,* the portion of the poor might be made more tolerable.[9]

Darwin's apparent relationship to Malthus and the use he made of the notion of population increase seemed to confirm the 'scientific' importance of the laws of population. But it also led to a peculiar jump in logic. It could no longer be averred that the law of increase vitiated change or improvement. However it was firmly insisted upon that if population increase occurred among the working class the change could only be for the worse. This feeling gave eugenics its peculiar air of catastrophism. It was Galton's opinion that civilisation had been on the decline since the Greeks. They passed highest in his tests of intellectual prowess and their civilisation had died out. Much of this alarmism about the decline of civilisation was, however, largely a projection onto history of contemporary conflicts and tensions. Eugenics was able to integrate two aspects of late nineteenth-century culture — fear of working class disorder and discontent and the rise in their numbers.

II

How far within this essentially social paradigm was there the emergence of a genuinely scientific and 'Darwinian' theory. Some recent work on Galton regards him as anticipating modern genetics largely because of his rejection of Lamarckianism and his insistence on a discrete unit of inheritance. This was true, however, of a considerable number of late nineteenth-century scientists, who have the right, in this case, to claim equal importance.

Far from anticipating modern genetics, Galtonism relied heavily on an old theme of nineteenth-century social research. He belongs, in fact, to the tradition of 'cerebral physiology' which Comte praised and whose characteristic proponent was the phrenologist Gall. This school attempted to link moral and social behaviour with certain physical or physiological features in man. Moreover it also used the idea of statistical distribution — the other 'modern' element in Galton.

Quetelet's work illustrates this trend. His *Treatise of Man* (1835) was translated into English in 1842 by Robert Knox. It contained several suppositions about social behaviour. The first was that 'everything that pertains to the human species . . . belongs to the order of physical facts'. According to Quetelet,

Lavater has not hesitated to analyse the human passions by the inspection of the features and Gall has endeavoured to prove that we may arrive at similar results by inspecting the cranial protuberances (sic). There is an intimate relation between the physical and the moral nature of man, and the passions leave sensible traces on the instruments they put in continual action.[10]

Moral nature therefore produced a series of physical signs and the object of science was to see which signs related to different kinds of social and moral behaviour. Thus Quetelet tried to establish a relationship between racial character and the incidence of certain crimes, coming to the conclusion that,

the Pelasgian race, *scattered over the shores of the Mediterranean and in Corsica* is particularly addicted to crimes against persons: among the Germanic race which extends over Alsace, the duchy of the Lower Rhine, a part of Lorraine and the Low Countries . . . we have generally a great many crimes against property and persons . . . . Lastly the Celtic race appears the most moral of the three . . . .[10]

It was the notion of 'observability' which led to the statistical element in Quetelet. In this respect, Galton followed him. If social phenomena could actually be seen, then the social theorists must count and enumerate regularities. It was this notion of the transparency of social life which led to, what many of their contemporaries considered was, the gullibility of this school, particularly their tendency to accept certain popular superstitions as the material for investigation.

The third element was its concern with potentiality. The attraction of Gall to a popular audience was his indication of the inner connections between observable physical characters and personality. But another important element was the possibility of moulding character. If personality could be so easily observed it could be developed and latent potentiality drawn out. As writers on Gall's phrenology have seen, this brought some phrenological circles into close association with Owenism and the doctrine of the relation between environment and improvement.[11] In his researches Quetelet sought for the statistically 'average' man. By observing him he hoped to deduce general laws about human development. But in the search he noted, like Gall, the phenomenon of individuality and difference, with whose laws of development he was also fascinated.

Galton took up this tradition. Darwinism served in his work only to 'modernise' what even in the nineteenth century was regarded by many as an archaic pseudo-science of mind. He was an observer of the physical expression of moral laws. He noted individuality and he searched for the average. He was also concerned with potentiality. This search for potentiality rapidly became transformed into a coin with two sides — the discovery of 'genius' in some and the necessary support and freedom this required. But, on the other hand, the identification of criminality, madness and inferiority and its suppression or control. Lastly, he brought into this tradition a preoccupation with the actual laws of inheritance and with the experimental psychology which was developing in the nineteenth century. Upon his integration of these with the older tradition lies his chief claim to 'modernity'. At the same time he looked less for the 'explicitly' racial mixtures within nations with which Quetelet— in a tradition popular among certain writers in France — explained the social and political divergences within his own country.

At least two traditions derived from the school of cerebral physiology. One merged with existing physical anthropology. Physical anthropology was concerned with the identification of race. But this preoccupation became inter-twined not merely with physical classification but with the detection of other mental and 'psychical' differences between the races. However the two kinds of investigation were often separated. John Beddoe, for example, conducted a series of racial investigations into nigrescence and cranial size in England and

Ireland in the 1880s with the objective of establishing significant physical difference in hair colour, eye colour and other
features.[12] But Beddoe, on the whole, did not extrapolate from
this to significant character differences. In the debate on
national deterioration in the early twentieth-century the
committee appointed to investigate the question made reference to his work and asked the Anthropological Institute of
Great Britain what the possibilities of a further survey were.
But the object was primarily to measure differences in physique.[12] In contrast both Galton and Quetelet attached 'psychical
qualities to the measurement of physical difference.

But there was an overlap. Galton in the anthropometric
laboratory he established at South Kensington, investigated
new and finer techniques of measuring physical and mental
response. His researches were made available to the Anthropological Institute and the first organised anthropological
field trip, made to the Torres Straits in 1896, carried with them
anthropometric techniques Galton had pioneered.

They also carried with them the notion that physical difference measured moral and intellectual nature. In fact the conviction that moral and cultural differences left observable
traces was expressed in a book, *The Study of Man*, which
Haddon — the leader of the Expedition — published in 1898.
According to Haddon — who had worked in the Anthropometric Laboratory in Dublin — there was a distinct
linkage of culture and physical feature. The negro was prognathous and the white man orthognathous. Haddon's conclusion was,

that the decrease of the action of jaw muscles is concomitant with rise in
culture, that is to increased mental activity . . . . Thus culture may act two
ways on the skull, directly by enlarging the volume of the brain and
therefore increasing the size of the skull; and indirectly by causing a reduction of the jaw which reacts again upon the skull. One is not surprised then,
to find that the higher races have, as a rule, a greater breadth in the anterior
temporal region of the skull than the lower races.[13]

Strangely, it was Galton's disciple Pearson who demonstrated
by statistical tests that cranial measurements were largely
meaningless. But nonetheless the Galtonian school pushed the
physical explanation of culture to its limit. This had tended to

be obscured by their more 'Darwinian' aspects but their Darwinism was in fact founded on the basis of this much older tradition of physical measurement.

John Stuart Mill had attacked schools of this sort — as had the psychologist Alexander Bain — for reducing the problems of 'mind' to those of biology and physiology.[14] Galton was referring to this when he wrote in the introduction to *Hereditary Genius* that 'I am aware that my views which were first published four years ago in *MacMillan's Magazine* . . . are in contradiction to general opinion.'[15] By general opinion he meant academic psychology and philosophy. Bain and Mill regarded 'cerebral physiology' as usurping problems which belonged properly to philosophy and, to some extent, introspective psychology. This did not mean that they denied the obvious advance in the investigation of the physiology of psychological process. But they witnessed an attempt to resolve every problem of social and moral behaviour into physical measurement and this, they believed was ultimately futile. But in fact the strain of 'cerebral physiology' remained surprisingly strong in eugenics. Galton himself searched for a correlation between brainweight and insanity, racial type and finger prints and the physiognomy of the criminal type. Pearson, Galton's successor at the Anthropometric Laboratory and director of the foundation set up by Galton to pursue research into eugenics, though committed to a more rigorous statistical investigation of inheritance, retained a similar conviction. In 1903 Charles S Myers, the psychologist, quoted with approval an investigation by Pearson which attempted to discover,

whether dark-eyed or tall women bear more children than fair-eyed or short, whether brothers are more alike than sisters, whether inheritance is more marked in female or male.[16]

In 1906, after Mendelism had been introduced in Britain, Pearson investigated the relationship between intelligence in children and, among other things, the curliness and colour of hair. The results included the information that, 'red-haired boys are most numerous among the *Slow Intelligent,* while red-haired girls have a reversed heteroclisy, being most frequent among *Quick Intelligent* or the very *Dull*'.[17]

With regard to these investigations, Pearson admitted that,

It may possibly be hinted that these results are of little significance . . . . It may be so but much of science is the verification or refutation of impressions or opinions.[17]

This brings us very close to the intellectual character of eugenics. It was based upon the mathematisation of often a very vague conviction that significant correlations must occur in nature and above all, as Pearson put it,

the conception that underlying every psychical state there is a physical state, and from that conception follows at once the conclusion that there must be a close association between the succession or the recurrence of certain psychical states, which is what we judge mental and moral characteristics by, and and underlying physical confirmation of it, be it of brain or liver.'[18]

III

It also meant that of all social Darwinisms eugenics was the most easily put to work. It required only an opinion about a correlation and the opportunity for counting. Galton was able to leave considerable resources for eugenics in the form of a foundation, some of the funds of which were put at Pearson's disposal. The result was a Biometric Laboratory under Pearsons's direction which undertook eugenic research of the kind quoted above. In the early twentieth century its studies included investigations into alcoholism, tuberculosis, insanity and the compilations of human pedigrees in *The Treasury of Human Inheritance* in 1909.

The Biometric Laboratory also became the centre for the propaganda which Galton thought an essential part of eugenics. It fed information into the network of eugenic education societies which spread through the regions in the early twentieth century. Galton regarded this part of eugenics as essential. Instilling a consciousness of the importance of racial improvement was eugenically important in itself. It also led to the demand for state intervention in favour of eugenics. Eugenics, argued Galton,

must be introduced into the national conscience like a new religion. It has indeed strong claims to become an orthodox religious tenet of the future.[19]

In spite of Galton's confidence about the spread of eugenics there was always a strain of opposition to it among intellectuals. Firstly, the British psychological tradition was too closely involved with basic philosophical questions about mind to take seriously an emphasis upon total hereditary determination of behaviour. Even psychologists who investigated the physiology of mental perception confined their attention to simple mental processes and left the complexity of social and moral behaviour alone for the time being. Or, if they speculated about it tended to return to a more philosophical tradition. More significantly, social thought was based upon the notion of a degree of free will and educability in man. An important tradition of social Darwinism rested upon this basis. Thus Leslie Stephen in 1896 wrote in contemptuous tones of 'Some men of science (who) have endeavoured to show that genius or criminality is hereditary, and that if one man writes a great poem and another picks a pocket, it is always in virtue of their hereditary endowment.'[20] This view of evolution, in his opinion, denied that social institutions embodied creative free will. It also reduced the capacity for moral self realisation and rationality.

Similarly, Benjamin Kidd criticised Galton's view of civilisation as,

so elementary that there was no place in it for moral standards, or for any of those problems of the responsibility of the individual for the universal which have distracted the human mind since the dawn of knowledge . . . .[21]

But this was not merely the view of a section of intellectuals. Eugenics had a strong influence on the legislation of the period. Even so that legislation also embodied another approach to social life. This approach refused to reduce social problems to heredity. To do so was to totally deny moral responsibility to the individuals involved in society. Thus if all poverty were a product of heredity then how could one encourage the poor to take the kind of moral responsibility urged on them by, among others, the Charity Organisation Society? Alternatively, how could they be educated in socially desirable behaviour?

The Committee on Physical Deterioration, for example, considered the problem of child neglect as one susceptible to improvement by education. In their opinion,

many testify to the willingness of the poor to learn and a tractable disposition in contact with judiciously tendered advice.[22]

An explanation of child neglect by reference to hereditary inferiority suggested a quite different approach to the 'social problem' one which would necessitate administrative control rather than education.

IV

Eugenics nonetheless had considerable appeal. There was an area of eugenics with a 'radical' tinge to it. Part of this radical appeal was undoubtedly the fact that it was an arena where sex could be discussed. Discussion of the choice of mate and control of population was part of the social responsibility eugenics enjoined on the citizen.

So strongly was eugenic duty in this area urged upon the 'better stocks' that G. K. Chesterton was able to satirise the eugenic approach to marriage when he claimed that,

The most serious sociologists, the most stately professors of eugenics calmly propose that 'for the good of the race' people should be forcibly married to each other by the police.[23]

Although the type of eugenist described by Chesterton was well represented in the movement, eugenic discussion of sex acquired a slightly bohemian air. It suggested that the process of choosing a mate should be liberated from social conventions and false prudery and subjected to certain 'rational' criteria. The next step was to turn the light of rationality upon the process of sex itself and the roles of men and women. Pearson in his youth had belonged to a society to encourage the freer social mixing of the sexes and had moved in similar circles to those of Havelock Ellis. Havelock Ellis himself was interested, like Galton, in the anthropometry of the criminal type and in race degeneration. He moved from these interests to a more direct concern with the question of sex.[24] Marie

Stopes was also led to sexology down a similar path.

If a mate should be chosen according to eugenic worth this suggested a greater degree of freedom for women to choose in marriage. This, after all, had good Darwinian credentials. Darwin had argued that many secondary characteristics of animals were developed by sexual selection, and sexual selection was the prerogative, in nature, of the female. Those eugenists who appealed for greater female emancipation in matters of sex could argue that by releasing the choice of the female from economic and social constraints they were also releasing a major force in eugenic improvement. A. R. Wallace, for example, believed that women's choice was the best selective agency and that this would be achieved when, 'a social system renders all women economically and socially free to choose'.[25]

Opinions differed on how far this involved a complete transformation of sexual relations. Galton had merely argued for state bursaries to allow the poor scholar to marry young— a scheme which struck a chord in the heart of the novelist George Gissing. The inability to provide certain financial guarantees was a major obstacle for marriage for the poorer professional man — a dilemma discussed in a number of Gissing's novels.[26] Galton argued for a high social value to be placed on a good genealogy, which would, presumably, influence the choice of marriage partner. In *The New Machiavelli* (1911), however, H. G. Well's hero Lewington takes eugenic marriage reform much further, arguing for a state supported motherhood. This would allow the woman, freed from economic constraint, to choose her natural mates and to terminate other experimental relationships without social disgrace and economic ruin. Both Imperial glory and the bohemian ethic could be served in this way.

However, certain people pushed the argument far enough to outrage the more conventionally minded eugenists. Shaw, who supported the idea of state aid for mothers, was involved in a controversy in the *Pall Mall Gazette* of 1907 with C. W. Saleeby— a prominent eugenist— over the question of eugenics and sexual reform. Saleeby undoubtedly represented the majority viewpoint among the movement when he reaffirmed the necessity of monogamy and argued against the intervention of the state in affairs of the family. Saleeby argued that,

The work of Professor Westermarck . . . has shown that this primitive promiscuity never existed . . . there is no society on earth, however rude, that does not punish the unfaithful wife.[27]

State intervention would merely weaken the family and the function for which nature intended it, which was the responsibility for the protection and rearing of young.

The second 'radical' element was the eugenic attitude to the state. For example, although both Spencer and Galton believed that the differential birth rate affected society adversely, Galton's attitude to state intervention and social reform was not Spencerian laissez faire. This was even more so in the case of Pearson who came back from Germany in the 1880s committed to a form of Bismarckian socialism. Eugenics could support a number of arguments for state intervention for it allowed the state, at the very least, the right to secure better racial stocks and discourage the bad ones.

This, combined with the role it gave in social progress to an intellectual elite, made it highly attractive to the Fabians. Sidney Webb, for example, went to some trouble to harness the schemes of a section of eugenics to Fabian socialism. In 1909 he addressed the Eugenics Society to argue that the proposals of the Minority Report on the Poor Law (1909) were based on eugenic principles. He pointed to the policy of segregation of different classes of paupers — for example, the able-bodied from the sick, the insane from the sane. The Minority Report was, like the Eugenics Society, against laissez-faire in principle and it also wanted to prevent a return to 'sentimental private charity'.[28] Finally, eugenics demanded the health and welfare of citizens. It did not want poor relief to become the resort of the 'no hopers' but of the healthy but unfortunate citizen. According to Webb,

The Minority Report is drawn on strictly eugenic lines. Its authors claim . . . that it contains absolutely no recommendation contrary to true eugenic principles. It was indeed drafted in constant consideration of the eugenic argument, and it may therefore be considered actually as an outcome of the educational work of this Society.[28]

Although this was, to some extent, a case of special plead-

ing, the convergence of radicalism and eugenics was not confined to this example. A year later H. J. Laski was defending
the principles of eugenics, on the grounds that,

We are now . . . fostering the weaker part of mankind until its numbers have
become a positive danger to the community.[29]

Laski combined this argument with a plea for a minimum
wage and the regulation of hours of employment. He argued
against Pearson's view that child labour, by increasing the
economic value of the child, was good eugenic policy. It
shows, however, how the concept of the deserving and unde-
serving poor became written into a certain section of 'collectivist' thought though it re-emerged in a 'scientific' form
rather than a moral one. In that respect a section of socialist
opinion could easily find a home in eugenics.

But this alliance of eugenics and collectivism was of only
limited efficacy. Eugenics expanded into regionally organised
education societies. It spread its net over politicians, medical
practitioners and ordinary professional people. As it did so it
came more and more to represent the common sense of the
middle classes. This common sense reasserted itself over the
vague bohemianism and the *soi-disant* radicalism. As time
progressed it became clearer that eugenics meant the preservation and increase of the middle classes. Wallace noticed this. In
1901 he wrote to a correspondent who had advocated the
segregation of the unfit,

You seem to assume that we can say definitely who are the 'fit' and who the
'unfit' . . . . I deny this, except in the most extreme cases . . . . I believe that,
even now, the race is mostly recruited by the *more* fit — that is the upper
working classes and the lower middle classes.[30]

Wallace was, in the last part of his sentence, trailing his coat,
for as L. T. Hobhouse pointed out,

To read a good deal of what is written on this subject (eugenics) one might
suppose that the whole question is as simple as daylight. Often it would
seem as if the actual position of classes in society was taken as a measure of
their worth. Thus we hear a great deal of the relative sterility of the richer
classes and the fertility of the poorer, as if this were, in itself, sufficient
evidence of the multiplication of the unift.[31]

It was clear that the notion of a residuum of talent at the bottom of society awaiting liberation through eugenic policies was being replaced with an assumption that the social hierarchy reflected the natural one. Pearson, for example, was arguing in 1909 that,

If we look upon society as an organic whole, we must assume that class distinctions are not entirely illusory; that certain families pursue definite occupations, because they have a more or less specialised aptitude for them. In a rough sort of way we may safely assume that the industrial classes are not on the average as intelligent as the professional classes and that the distinction is not entirely one of education.[32]

Thus the Eugenics Society reacted to Webb's defence of the Minority Report with considerable hostility. Largely this was because it suggested that some unemployment at least might not be caused by heredity or the inferior faculties of the unemployed. The Eugenics Society conducted their own investigation and came to the conclusion that 'pauperism is due to inherent defects which are hereditarily transmitted'. These hereditary defects it listed as,

drunkenness, theft, persistent laziness, a tubercule diathesis, mental deficiency, deliberate moral obliquity, or general weakness of character manifested by want of initiative or energy or stamina and an inclination to attribute their misfortune to their own too great generosity or too great goodness, and generally to bad luck. Inquiry into nature of the bad luck or too great generosity usually resolved the matter into one of stupidity or folly on the part of the complaining victim.[33]

Views of this sort suggested an emerging alliance between eugenics and Spencerian individualism. Certainly eugenics became an additional part of the argument against social welfare. The contemporary debate on poverty and social reform was influenced by it.

Alfred Marshall, in the section in *Principles of Economics* on raising the standard of living of the poor, argued that social welfare might lead to,

Some partial arrest of that selective influence of struggle and competition which, in the earlier stages of civilisation, caused those who were strongest and most vigorous to leave the largest progeny behind them and to which more than any other single cause, the progress of the human race is due.[34]

Part of Pigou's *Economics of Welfare* was framed with these arguments in mind.

> Economists, it is said, in discussing, as I have done, the direct effect of the state of the national dividend upon welfare are wasting their energies. The direct effect is of no significance it is only the indirect effect upon the size of the families of good and bad stocks respectively that really matters. For every form of welfare depends ultimately on something much more fundamental than economic arrangements, namely the general forces governing biological selection.[35]

Pigou got round this by arguing that 'My reply is that the environment of one generation *can* produce a lasting effect because it can affect the environments of future generations. Environments in short as well as people have children',[35] a phrase that was then incorporated by Hobhouse — a strong anti-eugenist — into an examination paper at the London School of Economics.

But eugenics was not another form of Spencerianism. It was much more involved in social administration and much less hostile to certain forms of government intervention. Galton had divided eugenics into negative eugenics — the prevention of racial deterioration — and positive eugenics — the active encouragement of racial improvement. Both aspects required a degree of administration and control. Thus eugenists saw their role from the beginning as active campaigners for this or that measure.

The scare over 'National Efficiency' after the Boer War which had revealed the poor physical quality of army recruits was one such area of policy in which eugenics attempted to intervene. Pearson's Huxley Lecture of 1903 which explained national decline in terms of the differential birth rate was quoted by the Commission members appointed to look into the question of physical deterioration and the witnesses were asked to comment on its relevance to the problem. But their interest in eugenics went further than this. The Committee on Physical Deterioration (1904) approached the Anthropological Society of Great Britain for their advice about an anthropometric survey of the British Isles, and one of their final recommendations was the creation of a permanent body to measure the physical qualities of the British population by the methods and techniques essentially pioneered by Galton.[36]

Their interest in eugenics was partly because eugenics was a science of categorisation and so was social administration. Moreover, social administration was highly respectful of the social order. Poverty then as now was the main route for becoming the subject of administrative control. Wealth enabled an individual to escape it. Eugenics suggested that the poor — who fell most often under administrative control — were also the category which evolutionary science demanded should be brought under restraint in various ways. Eugenics also suggested a principle of treatment. This was the notion of segregating the undesirable in order to prevent them from bearing children. Eugenists exercised a great deal of pressure to obtain this aim when the Commission on the Care and Control of the Feebleminded sat in 1908. But segregation was also a policy applied to the poor, and as well as a practical aspect it also had a punitive one. Many eugenists were prepared — as traditional liberals feared and suspected — to employ the wholescale apparatus of the state to fine, chase and lock up individuals who could be said, at the very most, to be guilty of being unfortunate and, at the least, of being a victim of mistaken notions of heredity current in eugenic literature.

For a dislocation was occurring between eugenics and theories of heredity. As time progressed its foundations in the science of heredity looked less not more sure. Eugenics did not seem to be able to produce satisfactory explanations of heredity. This was reflected in the Committees. The Commission on the Feebleminded complained that,

two opposing doctrines have been submitted to us (1) According to the one, mental defect is spontaneous in its beginnings and has a great tendency to recur in descendants and thus is truly inborn. . . . (2) According to the other, the evil influences of the environment are far more important than the innate and spontaneous defect of mental capacity transmissable by inheritance.[37]

The Committee recommended, in fact, that current medical research should be directed towards elucidating the study of heredity. This was an admission that eugenics had in fact failed to totally convince.

The Committee also revealed the sharp division which had arisen among social theorists about the relevance of eugenics.

John Gray, secretary to the Anthropological Institute, had considerable confidence in anthropometry and believed that a certificate of anthropometric worth should be given to those aliens allowed to enter Britain. He also agreed with Pearson in his evidence to the Committee that,

Anything which decreases the difference between the birth rate and the death rate among the superior classes and increases this difference among the lower classes tends to produce a progressive deterioration of the average national physique.[38]

Professor Cunningham, also from the Anthropological Institute, disagreed.

. . . it is stocks not classes which breed men of intellect. These intellectual stocks are found in all classes high and low. No class can claim intellect as its special prerogative.[38]

V

Overlying these disagreements was the conflict in science over theories of heredity. Eugenics was brought into sharp conflict with Mendelism. Each operated on distinct theoretical premises. Pearson's view of evolution was posited, like Darwin's view, on small graduated variations. Bateson tended to see Mendelism as a justification for his view that evolution took place by large mutations and by laws quite distinct from those worked out by the biometricians.

But perhaps more vital was the impact of Mendelism on the programme of social reform proposed by eugenics. Mendelism suggested that heredity, in a number of ways, was more complex than the eugenists had allowed for. First of all Mendelism distinguished between qualities which were hereditary — genotypical — and those which were not truly hereditary — phenotypical. This distinction eugenics had tended to ignore by insisting upon the all-pervasiveness of heredity and of the inheritability of any variation that appeared in the lifetime of an individual. Thus, in Galton's view, all observable physical characteristics were inheritable or at least must be presumed to be so. In contrast Mendelism insisted on the demonstration of inheritability. It provided a set of protocols for doing so — law of segregation and dominance — at odds with Pearson's biometric techniques and which made confident social prophecy more difficult. Thus the eugenics movement which had averred that anything from criminality to alcoholism was hereditary found itself challenged to demonstrate that these tendencies followed Mendelian laws. This uncertainty would imply, at the least, a degree of hesitation about giving advice on problems of social policy.

This was, to a great extent, recognised. C. W. Saleeby, for example, admitted in 1914 that,

at the end of this first decade of modern eugenics we know very much less than many thought we knew in 1904, that a great many points on which Galton himself would have said, 'Here we have knowledge, here we may advise society', are points on which further knowledge, is now, ten years later, required . . . .[39]

It is true that Saleeby's sympathetic reception of Mendelism was, in part, due to the capacities for 'mysticism' it seemed at that stage to have. Nonetheless the conclusion he drew was true. C. B. Davenport the American Mendelian who was sympathetic to certain aspects of eugenics nonetheless felt obliged to take the eugenic school to task for certain of their theories of heredity. Pearson, however, retained a stubborn conviction of the rightness of his original approach. He took the hiatus produced in eugenic knowledge by the advent of genetics as evidence of a general attack on the notion of the importance of heredity. In particular he singled out the theory of the recessive gene as a Mendelian 'denigration' of eugenic values since it attributed a less *immediate* effect of heredity on the next generation. In reply to Davenport, he wrote that,

Mendelism is being applied wholly prematurely to anthropological and social problems in order to deduce rules as to disease and pathological states which have serious social bearing . . . on the basis of Mendelism theory it is asserted that both normal and abnormal members of insane stocks may, without risk to future offspring, marry members of healthy stocks.[40]

Pearson stuck to his original statistical conceptions of heredity into the 1930s and also attempted, according to Bateson, to retard the advance of Mendelian research.[41]

Nonetheless in two areas integration of biometric and genetic research occurred. Firstly, Pearson's use of statistical techniques in investigating heredity indirectly helped population genetics. Secondly, the ideological barrier which he had erected between genetics and eugenics did not last. Bateson was not hostile to certain social interpretations of biology. Nor did he believe that the notion of 'improvement' of stocks was an unworthy ideal. In fact his social views were similar to those of many eugenists.

The essential difference between the ideals of democracy and those which biological observation teaches us to be sound, is this; democracy regards class distinctions as evil; we perceive them to be essential. It is the hetrogeneity of modern man which has given him his control of the forces of nature. The maintenance of that hetrogeneity, that differentiation of members, is a condition of progress. The aim of social reform must not be to abolish class, but to provide that each individual shall so far as is possible get into the right class, stay there and usually his children after him.[42]

Thus, what ultimately, seemed to differentiate the eugenic from the anti-eugenic school was not so much one's views on heredity as those on culture. There were those who believed that culture was irreducible to heredity — without necessarily denying all connection between the two — and those to whom culture was ultimately heredity. This was not a difference between biologists and sociologists. A number of biologists, such as E. Ray Lankester and J. A. Thomson, would have subscribed to the view that culture could not be considered a branch of biology. Similarly a number of social theorists believed that certain cultural phenomena were reducible to biological law. Even Hobhouse who felt apprehensive about the intrusion of eugenics into sociology, believing that it threatened the existence of sociology as an independent discipline, nonetheless formulated his theory of the evolution of the rational in terms of biological order. Thus many social theorists who were concerned with the development of the theory and techniques peculiar to their own profession nonetheless kept an area, even if a minor one, reserved for the biological and often the eugenic creed.

When Westermarck in 1891 published *The History of Human Marriage* — a theory of the evolution of marriage by reference to the natural selection of instinctive behaviour in animals and man — Robertson Smith described his position as one in which, 'the laws of society are, at bottom, merely formulated instincts. . .'[43] A. R. Wallace leapt to his defence. According to Wallace,

> To Naturalists the very objections of Professor R. Smith that you assume the laws of society (as regards marriage) to be mere 'formulated instincts' will seem the greatest merit of your book.[44]

Throughout most of the late nineteenth century Wallace's opinion was shared by some social theorists and rejected by some naturalists. Wallace whilst rejecting eugenics looked to another biological foundation for social laws. In the early twentieth century this was increasingly found in the mechanism which Westermarck described in *The History of Human Marriage* — the character of instinctive behaviour.

VII RATIONALITY AND IRRATIONALITY

IN 1894 Benjamin Kidd published *Social Evolution* sparking off a widespread debate among social Darwinist thinkers. The novelty of *Social Evolution* was not its use of the analogy of natural selection to explain social development. It was its use of these ideas in the context of a theory of the importance of non-rational factors in evolution. Kidd's non-rational factor was religious belief. This, he argued, had proved immensely important in evolutionary survival. Religious belief increased the solidarity of societies and led to social efficiency. In fact, according to Kidd, reason was a socially disintegrating force since, rationally, man ought to follow self interest. But it was only by the sacrifice of self interest that social progress was possible and this needed the sanction of a religious and supernatural metaphysic. These non-rational codes protected society from the individualism which led to a rampant capitalism on the one hand and socialism on the other — both Kidd regarded as the expression of rational self interest. The non-rational factor in Kidd's opinion provided,

The integrating principle . . . providing a sanction for social conduct which is always of necessity ultra-rational and the function of which is to secure in the stress of evolution, the continual subordination of the interests of the individual to the interests of the longer lived social organism to which they belong.[1]

A vein of piety and of scientism in British life assured immediate success for Kidd's book. However the reaction against it was also considerable. Kidd had, in the words of the *Spectator* reviewing a later book, *Principles of Western Civilisation* (1902),

treated Mill and Huxley and Spencer not merely as wrong but hopelessly antiquated, *vieux jeu*, blind and obscurantist . . . . There is nothing the world enjoys so much as seeing the party which believes itself to have the monopoly of progress proved to be the party of stagnation and old instincts, and even prejudices given philosophic rehabilitation.[2]

The 'party of progress' had pioneered two things: the idea that Darwinism had shown that morality did not need the sanction of religious belief and, second, that evolution was a growth in rationality. In secularist circles therefore Kidd's book was regarded as anathema. According to the secularist H. M. Cecil it was the revival of irrationalism in a new guise. Huxley thought it 'clever from a literary point of view and worthless from any other'. Stephen called it a theory of the necessity for 'a little stupidity' in social life.[3]

D. G. Ritchie also leapt to the defence of rational evolution. In a review of *Social Evolution* he restated his belief that there was,

a social instinct independent of religious sanctions (it exists in germ among all gregarious animals), and many kinds of religion and many elements of religion frequently conflict with this instinct . . . . Rational considerations may support and supplement this social instinct and it is chiefly those forms of religion which are largely permeated by the direct or indirect effect of rational considerations that lend support to the social instincts of mankind.[4]

However, Kidd's book marks a watershed which separates one tradition of social Darwinism from another. It was the first of a series of reappraisals of the theory of the evolution of rationality made familiar by Stephen, Ritchie and Bagehot. This reappraisal was a rediscovery of the persistence in social life of irrationality, or of the less tendentious concept Kidd used, ultra-rational or non-rational conduct. It was also an attempt to explain its survival by the use of Darwinian analogies.

Various factors encouraged this. Politics had by the turn of the century developed in ways unacceptable to those who would, by inclination, have belonged to the tradition of liberal rationalism. The aim of integrating the working class into a moral consensus and liberal political tradition had not been achieved to the extent that many liberals had hoped. The kind of society which democratic popular politics had produced was not the kind of society which liberal intellectuals had hoped for. In concert with these developments occurred an increasing resort to an explanation of political behaviour in terms of irrationality. J. A. Hobson believed the chauvinism associated with the Boer War was 'an instinctive display of some common factors of national character which is outside reason and belongs in ordinary times to the province of the subconscious'.[5] Graham Wallas's observations of electoral politics in the early 1900s led him to similar conclusions. Popular politics had brought into play the language of sentiment and emotional appeal.

If the word 'Wastrel' for instance, appears on the contents bills of the *Daily Mail* one morning, as a name for the Progressives during a County Council election, a passenger riding on an omnibus from Putney to the Bank will see it half-consciously at least a hundred times, and will have formed a fairly stable mental association by the end of the journey . . . . The contents bills, indeed, of the newspapers which were originally short and pithy . . . have developed in a way which threatens to turn our streets . . . into a psychological laboratory for the unconscious production of permanent associations.[6]

In Wallas's opinion political life in early twentieth-century Britain did not show much evidence of the evolution of reason. This did not necessarily lead to a repudiation of rationality or a return to belief in religion. It did however mean a re-examination of the utilitarian theory of action. Kidd suggested that utilitarianism was wrong for men acted not because they perceived some reasonable outcome to their action but out of superstitious regard for religious authority. Similarly Wallas argued that,

the connection between means and ends which they (actions) exhibit is the result not of any contrivance by the actor but of the survival in the past of the 'fittest' of many varying tendencies to act.[6]

Those whose unintentional behaviour secured the welfare of society were socially selected. Thus rational man had less evolutionary importance than Bagehot and Stephen believed. Knowledge of the character of social evolution was no guarantee of a better and more secure social order.

The object of many social theorists became therefore to find out the origin and basis of desirable social behaviour. If it could not be attributed to man's reasonable perception of the good, to what evolutionary characteristic could it be attributed? In the case of some of these social theorists they refused to jettison rationality altogether but sought instead a theory of social action which gave it less importance. Increasingly the concept of instinct was used for these purposes.

The growing influence of the concept of instinct can be traced in the late nineteenth and early twentieth century; in anthropology in the writings of Edward Westermarck and W. H. R. Rivers; in social psychology, in the works of William McDougall who by the early years of the twentieth century had established himself as one of Britain's leading social psychologists; and in political theory, in the works of Graham Wallas and W. Trotter. Wallas, Rivers and Westermarck taught on the first courses in the social sciences at the LSE. It was largely due to them that the concept played a significant part in the development of the discipline.

The theory of instinct concerned the biological determination of both man's nature and his social behaviour. W. H. R. Rivers defined instinct as 'an inherited disposition to behaviour'.[7] Darwin was considered to have pioneered the 'modern' study of human instinctive behaviour and to have provided social theory with the requisite concepts for its analysis. For example, Westermarck claimed to have,

made myself conversant with the main teachings of Darwinism and had found the theory of natural selection, especially in its application to instinct, to be of enormous importance in the solution of the problems with which I was occupied.[7]

Darwin initiated a number of ideas and theories relevant to the use of instinct in describing social behaviour. In the first place, he applied it to human behaviour. This was a considerable development for previously, although natural history had granted that a residue of instinctual behaviour existed in man, it had also considered that the presence of instinct varied inversely with intelligence. Consequently, although it was important in describing animal behaviour, its influence upon human activity was considered to be weak and relatively unimportant. Darwin, however, argued that the operation of instinct in human behaviour was significant and that instinct and intelligence were not necessarily opposed. Instinctive behaviour was generally accompanied by 'a little dose . . . of judgment or reason'.[8] Moreover, although

Cuvier maintained that instinct and intelligence stand in inverse ratio to each other . . . Pouchet, in an interesting essay, has shewn that no such inverse ratio really exists. Those insects which possess the most wonderful instincts are certainly the most intelligent.[8]

Darwin also had applied instinct to describe social behaviour, for example the cell making activity of bees — a form of co-operative behaviour— and in the Appendix to the *Origin* he wrote on the instinct of bird migration. In the case of man he tended to restrict the workings of instinct to categories such as 'self-preservation, sexual love, the love of the mother for her new-born offspring, the power possessed by the latter of sucking and so forth'.[8] But he also suggested in the *Descent* that there existed a 'social' instinct for co-operation, from which might be derived other, more complex social behaviour, such as moral feeling. Finally, the theory of natural selection provided a means of linking the survival of instinct in man to essential social functions. Darwin was doubtful about whether all instinct was 'functional' for society. However, his doubts about this tended to be suppressed in the work of writers who followed him.

One characteristic of instinct was the guarantee it seemed to give that certain socially desirable behaviour would continue in a world of shifting values. E. A. Westermarck, in his book *The History of Human Marriage* (1891) used the theory of instinct to argue that the pairing family was present in all historical periods. This was proved, in his opinion, by the structure of primate families. Underlying this structure was an instinct which reasserted itself in human society. Its survival arose from the importance for the protection of the young of a species of the pairing bond between parents. Westermarck tended to reduce the varieties of marital form found in historical periods to a single one— the pairing family— of which the varieties were all epiphenomenal forms.[9]

A similar comforting reassurance existed in the notion that an acquisitive instinct underlay the institution of property. William McDougall, the social psychologist, listed the acquisitive instinct as one of the major characteristics of both animal and man. In an influential book *Property; its Origin and Development* (1892) — on the syllabus of the London School of Economics sociology course— Letourneau described the relation between property and instinct. According to this 'the instinct of property is but one of the manifestations of the most primordial of needs, the need of existing and securing existence to offspring'.[10] This instinct led to the struggle for territory and food in animals and man. It threw 'a singular light upon the origin of the right and instinct of property'.[10]

But a section of anthropologists hesitated before applying the notion. Anthropology had been associated with an investigation of cultural variety and the theory of instinct seemed to suggest that cultural difference had considerably less importance than had been assumed. Robertson Smith criticised Westermarck's theory of human marriage for this reason. Similarly W. H. R. Rivers, although he used the notion widely, tried to find some means of reconciling it with cultural variability. For example, how could the notion of the connection between property and the acquisitive instinct be reconciled with the apparent 'communism' of Melanesian society? Rivers' solution was to argue that 'the highly diverse varieties of mankind may have acquired different instincts'[11] by, for example, the process of inheritance of acquired characteristics. This meant that the possibility that some form of socialist society might exist in the future was not necessarily vitiated by the existence of acquisitiveness. On the contrary, as new institutions developed so would new instinctual forms of behaviour adapted to them.

This had a political consequence. It meant in Rivers' opinion that,

Since the aspect of socialism which attracts most interest is that connected with property and the management of property, the instinct involved in this case will be the instinct of acquisition, and the first step in the process of discovering whether from this point of view socialism is contrary to human nature would be to discover whether such an instinct of acquisition exists.[11]

Rivers in fact found that a countervailing instinct to that of acquisitiveness existed. This was the herd instinct and upon it, Rivers argued, the possibility of communistic societies depended.

However, in spite of these modifications, the theory of instinct acquired a highly conservative bent. It seemed to suggest that there was a scientific basis for a metaphysics of 'human nature'. In McDougall's view it supported the idea that there existed certain human characteristics, extremely stable and persistent, and that if the experience of each generation was in any manner or degree transmitted as modifications of the racial qualities, it was only in a very slight degree. So that a moulding effect was achieved only very slowly and in the course of many generations. Upon these characteristics rested the persistence of institutions such as the family, property, nations and war.

What then was the attitude to this concept of the theorists who believed in the possibility of social progress and in the importance of reason in social life? William McDougall carried the notion of instinct further than most social theorists. In his opinion all or nearly all of the social behaviour of man had some instinctual basis. He did however make two provisos. Firstly, instinctual patterns, whilst they are similar in all races of man, can be satisfied by different social institutions. This explained cultural diversity. Secondly, McDougall argued that in the course of evolution instinct is modified by intelligence.

This final premise is important. It was the means by which liberal theorists could defend at least some of their propositions about social progress. Both McDougall and Wallas attributed the influence of this idea to Hobhouse's *Mind in Evolution* (1901) and McDougall dedicated his book *The Group Mind* (1920) to Hobhouse. McDougall argued, for example, that only in the earliest societies could instinct be found in its purest forms. Later societies modified instinct by education and moral teaching. In other words, instinct could be altered by the emergence of rationality. There was a higher stage of evolution,

in which conduct is regulated by an ideal of conduct that enables a man to act in the way that seems to him right regardless of the praise of his immediate social environment.[12]

For McDougall, however, instincts were still functional; they operated in contemporary society to produce socially useful behaviour and they formed part of ordered evolutionary development.

In this respect McDougall was closer to Hobhouse than

Wallas. Wallas thought that instinct and reason were much more opposed. In general he felt that instinct was an inherently conservative and anti-liberal force. He held the belief that the most favourable social system for its operation was that of mass democracy. Instinct was in general an expression of *collective* behaviour and collective behaviour was inevitably non-rational and anti-liberal. In this respect Wallas clearly showed how far a section of liberalism, whilst accepting the inevitability of democracy, retained certain reservations about its character.

W. Trotter, for example, in a series of articles for the *Sociological Review* in 1908 and 1909 and in *The Instinct of the Herd in Peace and War* (1916) elaborated the notion of an instinct of gregariousness in mankind. This, he argued, was revealed by the tendency to split into nations and groups within nations. Trotter's 'herd' was very similar to that of the French theorist Gustave Lebon. Gregariousness was a 'primitive' instinct accompanied by the fear of loneliness, liability to panic, suggestibility and a strong and irrational demand for group conformity. This led to a constant contradiction between the individual, the bearer of traditional liberal values, and the crowd. This struggle was between the emancipated and rational individual and the 'primitive' group. William McDougall was later to attempt to humanise the 'herd' in the Group Mind by attributing to group consciousness a kind of metaphysical unity which, in his hands, almost amounted to a religious or spiritual bond. This was a collective 'mind' rather than a herd instinct — a subtle change which marked an attempt to see the herd as a communion of racial souls rather than of degraded men and women.

Wallas and Trotter, on the contrary, argued that cultural evolution and instinct were opposed. The instinctual drives could obliterate or dominate the rational side of man's character. The result was a lapse into 'primitive' or 'savage' mentality. J. A. Hobson agreed with this and to some extent must be credited with the development of the idea. In the *Psychology of Jingoism* (1901) he had argued that, 'the hypothesis of reversion to a savage type of nature is distinctly profitable. The war-spirit as displayed in the non-combatant mass-mind is composed of just those qualities which differentiate savage from civilised man'.[13] In particular Wallas believed that the danger of atavism increased because of the direction of social evolution. The features of contemporary society which led to a revival of instinct were listed in *The Great Society* (1914) as urbanisation, the strains of industrial society, the character of mass politics including Imperialism. The contradiction between instinct and reason increased, Wallas believed, as evolution progressed. The only solution to this was a more realistic assessment by liberals of the possibility of social and political progress.

From being originally an explanation of certain forms of social behaviour, instinct was being transformed into a moral sermon on the future of civilisation. For most British political and social theorists the relation of instinct to civilisation provided either the reassurance of convervative stability or an unpleasant reminder of the primitive in man. There was very little application of the idea of instinct as a dynamic force in recreating social life or as a force to be welcomed as destroying certain unpleasant aspects of contemporary civilisation. It was unusual to hear the kind of opinion expressed by the German social theorist Scheler who predicted according to one recent commentator on his work that,

The power of the mind acquired through sublimation of instinctual energy was extremely limited however. Mind could not *prevent* the inevitable triumph of instinctual drive; it could only *delay* it.[14]

In British literary culture — which had always expressed greater hostility to industrial society — the notion of instinctual drive and its benevolent effects was much more readily accepted. Whereas Rivers regarded the emergence of instinct as the product of childhood, stress or madness, a vein of literature regarded its reassertion as evidence of the logic and symmetry of nature. D. H. Lawrence, for example, in a letter to Ernest Collinge in 1913, wrote that,

My great religion is a belief in the blood, the flesh as being wiser than the intellect. We can go wrong in our minds. But what our blood feels and believes is always true. The intellect is only a bit and bridle. What do I care about knowledge? All I want is to answer to my blood direct without fribbling intervention of mind, or moral or what not.[15]

Whereas Wallas saw, in the re-emergence of instinct due to the frustrations of industrial society, a dangerous atavism, a section of British literary culture regarded instinct's revenge on civilisation as both necessary and desirable.

The theory of instinct encompassed both McDougall's 'group mind' and Trotter's 'herd'. Both collectivities were assumed to be bound together by instinct but there was a subtle distinction between the two conceptions. The 'herd' was also Wallas's and Hobson's mob — a primitive and dangerous reversion to the savage. But to McDougall the group mind represented an almost mystical and religious union. Race and nationality became a matter not of observable physical and cultural difference but of spiritual unity and of some kind of historical memory. According to McDougall,

Nationhood is, then, essentially a psychological conception. To investigate the nature of national mind and character and to examine the conditions that render possible the formation of national mind and tend to consolidate national character, these are the crowning tasks of psychology.[16]

Whatever their view of the character of instinct most theorists were agreed on this point. They were also agreed that national feeling was built upon instinct. According to Arthur Keith, the anthropologist, speaking at the Boyle Lecture of 1919,

I have no wish to analyse the subconscious states and instinctive reactions which rule and bind together the members of a primitive community; what I want to make clear is that the tribal (or national) instincts serve not only as a machinery for binding the members of a community together, but also as a means of separating them from all surrounding groups. Within the community this machinery compels unity of sentiment and of action, it serves to repress schism and faction.[17]

They were also agreed that the character of such sentiments was either non-rational or to put it more strongly irrational. A question in the Sociology exam at London University in 1910 asked the students to discuss 'How far . . . recent psychological analysis has tended to circumscribe the efficacy of 'rational' factors in social life?'[18] Liberals put down the failure of rationality to the limitations of 'natural man'; socialists and anarchists to the corruption of institutions; conservatives merely thought that sentiment was, in any case, a better guide to political action than reason. In any case sentiment could be shown to have 'a function' in binding together classes and nations. The irrational might — as Kidd believed he had shown — reveal a greater evolutionary wisdom, one which transcended individual reason.

As a concept, instinct showed early on a capacity for inexactness. Did it, for example, describe all behaviour or only some; to what extent was consciousness or intelligence combined with instinctive actions; were all instincts useful in survival? Darwin in the Appendix on instinct which G. J. Romanes published in 1883 seemed to think that some mixture of intelligent perception and instinctive action was possible. He also observed certain patterns of animal behaviour which seemed to him to be residues from a previous use but which now had lost their utility for the animal concerned. He also believed instinct to be hereditary but, if instinct and intelligence operated together, did this imply that intelligence could, in some way, direct the heredity of animals?[19]

These questions perplexed the proponents of instinct. It became, as William Kirby in the *Bridgwater Treatise* wrote, 'a field in which whoever perambulates, may wander in "endless mazes lost" '.[19] G. J. Romanes saw the notion of intelligence in instinct as justification for introducing an element of Lamarckianism into evolution by asserting that the intelligent initiation of an action by an animal may very well be moulded into an inherited instinct. On the question of how far all behaviour was instinctive C. Lloyd Morgan tried to reintroduce a sharper distinction between intelligence and instinct largely to prevent instinct from assuming a completely dominant role in descriptions of animal behaviour and one which would be rendered largely vacuous by its extension to cover every aspect of behaviour. In his view by separating out the two strands — instinct and intelligence — some kind of hierarchy of human and animal behaviour might be constructed and a distinction between the two might be preserved.[19] His contemporary, the psychologist and anthropologist Charles S. Myers, however, refused to distinguish them. In his view,

Thus the psychology and physiology of instinct are inseparable from the psychology and physiology of intelligence. There is not one nervous apparatus for instinct and another for intelligence. We ought to speak not of instinct and intelligence but of instinct-intelligence, treating the two as one indivisible mental function which in the course of evolution has approached now nearer to so-called instinct, now nearer to so-called intelligence.[20]

By the early nineteenth century Bergson's views had further complicated the matter. He had introduced a wider category than intelligence–consciousness or intuition — into theories of instinct. Instinct, according to this, made use of a broader consciousness which was non-intelligent or a-intelligent. Finally the advent of Freud made things even more complex. Freud appeared to use the notion of instinct but in a way which puzzled its proponents in Britain. They admitted that a particular instinct might be satisfied by different institutions but not that an instinct could be as malleable as Freud proposed. McDougall, for example, considered that the sex instinct of which Freud wrote was indeed a valid concept. But for McDougall every institution had an instinctive basis. Sex might have one (or to speak more accurately marriage and procreation might have one) but McDougall could not see how Freud could specify only one instinct since there existed myriads of social institutions based on instinct.[21] McDougall and other theorists of instinct, at that time, had no notion of an unspecified instinct which might latch itself onto various objects of many kinds. In their quite rigid and moral and social world this would be equivalent to a kind of evolutionary anarchy.

But where British instinct theorists did find value in Freud was in the way the notion of instinct could be combined with that of the unconscious. Trotter, a friend of Ernest Jones, thought that the secret of the atavistic quality of instinct lay in a kind of historical memory which was very much equivalent in his eyes to the unconscious. This explained why the crowd for example had to receive certain signs, symbols and excitements before it would display instinctive behaviour. This conception was of course derivable from Lebon. Nonetheless Freud's work on the unconscious gave the concept of the historical memory an apparent legitimacy. W. H. R. Rivers (for whom any new idea had an irresistible attraction) used the notion of the unconscious and instinct in his investigations of the reaction of combatants in the Great War. The unconscious was a place where impressions and ancestral memories were stored. The cataclysm of the war had acted, in Rivers' view, as a stimulant to it.[22] Wallas and other writers on social psychology in the 1920s toyed with the notion of reconciling the writings of Freud with the view of a cultural evolutionary memory which mental disturbance in the individual could dredge to the surface and make operable.[23] This also applied to the group. Thus the collective mind, like the individual one, was considered to have its unconscious. As such the works of Freud became intermingled with a whole series of early twentieth-century pessimistic accounts of the failure of civilised man to escape a savage past.

To the animal ethologist however there was decreasing advantage in a concept so diffuse. On his arrival in America McDougall encountered the behaviourist backlash against his theory of instinct.[24] However, criticism of the theory of instinct was also present in England much earlier. In 1909 at a particularly disputatious meeting of the British Psychological Association, McDougall confessed himself,

disappointed to find that all of my colleagues assign to instinct a vanishingly small part in the development and operations of the human mind. They seem to be agreed in recognising the suckling, wailing and crawling of the infant as having some instinctive basis; but beyond this none of them is prepared to go. From this it follows that they must regard the whole of my book on 'Social Psychology' as purely fanciful . . . .[25]

The theory of instinct had in fact to be rescued from the innumerable attempts to make it bear the brunt of all social behaviour — to express the 'élan vital', the collective mind as well as the migration patterns of the swallow and the institution of property.

The disquiet expressed by McDougall was premature. There was agreement among a wide circle of psychologists and animal ethologists that the concept had been employed too loosely in the early twentieth century, but later on its rescue was accomplished by, among others, Lorenz and Tinbergen. This rescue was carried out by refining some of Darwin's original ideas. There was an increasing acceptance that animal behaviour was a complex structure of reflex, learned behaviour and instinct and sometimes one or more conflicting instincts.

Consequently less attempt was made to find a single instinctive drive to explain a particular behavioural pattern. There was less introspective analysis and more controlled observation and experiment. Along with this went a willingness to see instinct as, in some circumstances, maladaptive and, particularly important, as capable of redirection towards different paths and objects. But what persisted well into the twentieth century were the disagreements about its application to human history.

Instinct had become in the early twentieth century, and has remained a means of philosophising about civilisation, racial differences and the direction of evolution. It had also become a means of emphasising the irrational. Hobhouse felt this keenly as a theorist of the development of rationality. He witnessed in the early twentieth century what seemed to him to be an irrationalist reaction. In 1921 he wrote,

Psychology, which begins to reduce the play of mental activity to a science, has not fostered the conduct as a reasoned art. On the contrary its tendency is to emphasise the primacy of feeling, the sway of instinct, the prevalence of the irrational in the mass movements of mankind. What is still more remarkable, philosophy itself once the supposed guardian and advocate of reason shares in the irrationalist tendency. We shall end by defining man as the irrational animal and the modern philosopher as his prophet.[26]

He was referring partly to certain of the criticisms of utilitarianism. Mill, Ritchie and Darwin himself had seen associationism as a means of criticising intuitionism — the 'support of false doctrines and bad institutions' in Mill's words.[27] Darwin had historicised associationism. By doing so he had explained how with ordinary psychological materials — the organism's reaction to its environments and the impressions and associations it built up during this experience of the environment — complex behaviour might have evolved. Instinct formed one of these psychological properties. It was a means of perpetuating behaviour and a residue in the organism of the impressions built up by its reaction to its environment.

However, the liberals of the late nineteenth century took a step further. They saw in instinct a potential critique of utilitarianism. Stephen had claimed his theory of ethics to be 'a new armoury wherewith to encounter certain plausible objections of the so-called Intuitionists'[28] — primarily their objection to the notion that a natural/scientific explanation could be found for the development of the moral sense. But Stephen also turned the theory of instinct against Mill and Buckle insisting that the philosophical basis of their politics must be reconstituted. Mill and Buckle had, he argued, considered man at birth a *tabula rasa* on which experience wrote and hence that, 'every social bond like every logical principle is to be resolved into a case of arbitrary association'.[28] In contrast Stephen had argued that Darwin's discoveries had shown that there were, indeed, distinct psychological graduations in mankind corresponding to their institutions. He had done so without surrendering ground to intuitionism but mainly by developing the theory of instinct and graduated evolution. Therefore whilst, on the one hand,

No reformer who duly estimates the gigantic power of impregnable stupidity should underrate the value of this scepticism as at least a provisional frame of mind. Undoubtedly we are grievously inclined to regard transitory customs, political and intellectual, as part of the unalterable framework of the universe.[28]

On the other hand, the view that behaviour was purely a product of experience was described by Stephen as 'as arbitrary as that of the antagonistic theory'.[28]

This meant two things. There must be some notion about the innate propensities and instincts of individuals employed in describing society and social institutions. Different individuals would react in different ways depending on the hereditary material of which they were composed. This view became a popular culture among intellectuals of the late nineteenth and early twentieth century. It seemed the most obvious lesson of the kind of evolutionism associated with Darwin. All the theorists of instinct began their work with an explicit renunciation of utilitarianism and the *tabula rasa*.

However the second meaning was political. To some liberals like Wallas the existence of an inherited disposition called 'instinct' reduced the chances of educability and social reconstruction implied by previous liberal philosophies. To others like Stephen it meant a reassertion of inequality against the idea of human equality. It implied different instinctual dispositions and also, more important, different psychological make-ups in different social stages. Thus in Stephen's view,

When people said that the negro slaves were lazy because they were negroes, and therefore doomed by nature to a perpetuity of flogging, it was necessary to point out that the laziness might be due to the slavery as well as to the negrohood. But it is a gratuitous assumption that slavery explains the whole divergence and that a negro differs from a European only as a man in a black coat differs from one in a white. The assumption becomes even more grotesque in the case of sex. Yet Mill and his followers are apt to slide into such conclusions and to lay down the equality of man as positively as the most dogmatic of *a priori* metaphysicians.[28]

Negro 'inferiority' was not an expression of their environment but their environment an expression of their 'inferiority'.

A similar philosophical and political debate crystallised around the twentieth-century revival of instinct theories, with the behaviourist occupying the place previously accorded to the *tabula rasa* theorist. This occurred in spite of the fact that, on a practical level in the psychological sciences, behaviourism and the theory of instinct were not necessarily irreconcilable.[29] Darwin's psychology, in spite of his use of instinct, was associationist. It treated behaviour as the product of the interaction of the individual and the environment in a way perfectly compatible with certain of the formulations of behaviourism. Darwin regarded human instinctual drives as of the most general and basic kinds. He was also highly suspicious of too much reliance on the notion of the 'innate' in analysing behaviour.

However, discussions on instinct acquired, as Hobhouse saw, a philosophical bias and here the divisions are as deep as ever. Most of them revolve again around the question of the educability of human nature, the persistence of certain social institutions and the irrational in behaviour. However, modern theorists of the role of instinct in human affairs are marked by a strong emotionalism more reminiscent of Lawrence's appeal to the authenticity of human impulse than to Wallas's fear of the irrational. Partly through Lorenz's reinterpretations of animal society, the distinctions between the 'savage' and the 'reasonable' and the 'animal' and 'human' have been made less sharp so that human instinct has been rescued from the idea common to the early twentieth-century liberals that it represents a savage residue of man's past.[30] In many contemporary works it has become, instead, a trustworthy and valid guide to human conduct.

This reappraisal of instinct in society has been strongly linked to the rediscovery of the moral and emotional life. In a passage which strongly recalls Benjamin Kidd, Lorenz proclaims, 'in fact there is nothing left in civilised society which could prevent retrograde evolution except our *non rational sense of values*'.[30] Kidd hoped that an institution — organised religion — could be guaranteed survival by the depth of those non-rational values which supported it. In Lorenz, there is a similar insistence upon the importance of continuity and tradition in human culture. This applies to all institutions to which a strong emotional appeal is attached. What he and others who have recently taken up his ideas offer is an 'affective' sociology whose function it is to naturalise deeply held feelings about the social environment and social institutions. This is put perfectly by Mary Midgley in a recent book, 'Understanding the structure of human feeling should help us to determine how far different ways of life are possible.'[31] Ultimately, despite references to science and nature, this sociology proceeds by registering emotional reactions as a guide to social action. Lorenz's sociology relies entirely upon this. For him society exists to satisfy deep instinctive needs for sociability. Social problems are largely the individual frustrations we experience. Even international tension can be dissipated through an emotion of well being — friendship.

As B. F. Skinner said, in another context, the fact that our emotions are important to us — and it is by recognising our state of mind that our survival to a great extent depends — does not make them necessarily a good guide in any investigation.[32] This applies particularly to an investigation of human social behaviour. What distinguished theorists of instinct such as Hobson and Wallas was that they saw emotion in politics and social life frequently as the legitimisation of prejudice; to put it in Hobson's words as 'the politics of jingoism'.

However, although Hobson and Wallas, by labelling what they defined as instinctive behaviour as 'atavism' and as an evolutionary survival, hoped to convey their distaste for its manifestations; in another sense they had surrendered arguments to their enemies. They did so by removing the explanation of the origin of 'instinct' from the realm of existing social experience. It became a characteristic of man's natural past not his present day social life. However critical they were of instinct, their mode of explanation tended to accept it as a 'natural', persistent and largely ineradicable feature of human social existence. Therefore, despite their insistence upon the atavism of instinct, they can be justly regarded as precursors of contemporary uses of instinct in describing social behaviour.

Stephen's digressions on character, heredity and society point to an important swing away from the use of Darwinism to defend and promote liberalism and radical reform. It even signals the revival of the kind of intuitionism which the *Descent* had deliberately set out to destroy. The theory of instinct played an important part in this. Instinct began to have all the characteristics of intuition and by the time Bergson arrived it had even become indistinguishable from it. It was scientific and Darwinian, in the opinion of its protagonists, but arose from very much the same kinds of rhetoric about the immutability of human nature, the inevitability of certain institutions and the natural basis of inequality that had been characteristic of the theories which originally it had been intended to attack. The compromise which Stephen effected between history, evolution and heredity was upset in favour of the theory of an inner hereditary impulse in history which moved society towards inevitable conclusions. The attempt to give this political belief a scientific basis was done mainly through the concept of instinct.

A number of liberals had signalled in the 1860s and 1870s their break from Mill and classic liberalism. Bagehot thought the emotional attachment of the people to symbols of monarchy was good for social order. Stephen and Huxley began to consult their feelings, and those of their social circle, on the question of race and, in the process, rediscovered a validity in their feelings of superiority over the black man. This rediscovery of the importance of feelings was often the re-establishment of areas of social privilege which, it was believed, must be defended at all costs but for whose continued existence rational defences as opposed to emotional ones no longer existed. But it is only fair to point out that to others, like D. H. Lawrence, emotion signalled an area of privacy which had to be protected from the encroachment of a dehumanising industrial civilisation. To an even smaller minority, like Shaw, emotion and instinct provided the impulse from which civilisation itself could be reconstructed. It was often from Darwinism that the terms and concepts were borrowed to reconstruct political theories on these lines. One could argue, however, this produced not a scientific explanation of the irrational and emotional in social life but a 'naturalisation' of it.

In spite of the criticisms his successors made of Mill's view of 'human nature', he had seen that prejudice and emotion were influential in politics. He was inclined, however, to explain this by more prosaic factors — by the interest that individuals and social groups felt in protecting institutions and caste. He was not inclined to overvalue these emotions as the support of social order nor to regard the object of politics as their propitiation.

In his autobiography Mill considered that,

I have long felt that the prevailing tendency to regard all the marked distinctions of human character as innate, and in the main indelible, and to ignore the irresistible proofs that by far the greater part of these differences whether between individuals, races, or sexes are such as not only might but naturally would be produced by differences in circumstances, is one of the chief hindrances to the rational treatment of great social questions . . . .[33]

Mill mistook, as in so many other cases, the temper of his times and class. His belief had a short life in English intellectual culture. With a few exceptions, the opposite was asserted. Stephen represents a half-way house to a complete reconstruction of theories of innate inequality. These theories and the reasons for their dominance in intellectual life are the subject of the next chapter.

VIII RACE AND CLASS

THE advent of Darwinism in the 1860s took place when several debates about race and class were raging. In 1861 the American Civil War began. This raised the question of the abolition of negro slavery. In 1865–6 the Governor Eyre controversy split British political society. A revolt by West Indian negroes was put down by Governor Eyre who executed the leader of the rebellion without regard for the legal process of trial and judgment. Whether Governor Eyre should be prosecuted for this exercise of arbitrary power became a major political issue strongly dividing intellectuals. Mill, Darwin and Huxley were for his prosecution. Among others, Carlyle and Ruskin against.[1] These were, in part, racial issues, but not completely. Liberals of the 1860s felt that the question of arbitrary government and political and racial equality was indivisible. Their opinions on the black–white issue were influenced, to a certain extent, by others more directly related to these questions within Britain itself.

The 1860s saw the Liberal Party ally itself to the issue of the extension of working-class suffrage — a movement which led in 1867 to an extension of the vote to the better-off sections of the working class enacted by a Conservative government and following widespread agitation. The debates on race were, in this period, also debates about the question of equality in general and about political democracy in Britain.

On questions of race Darwin took a liberal position. In his account of the *Voyage of the Beagle* (1839) he included an appendix directed against the institution of slavery which he had encountered in his travels in the Spanish and Portugese colonies. He also recounted the clash he had with the pro-slavery Captain Fitz-Roy who commanded the *Beagle* and which almost led to Darwin's abandonment of the voyage. These opinions he retained. According to his son,

With respect to Governor Eyre's conduct in Jamaica, he felt strongly that J. S. Mill was right in prosecuting him. I remember one evening, at my Uncle's, we were talking on the subject, and as I happened to think it was too strong a measure to prosecute Governor Eyre for murder, I made some foolish remark about the prosecutors spending the surplus of the fund in a dinner. My father turned on me almost with fury, and told me, if those were my feelings I had better go back to Southampton; the inhabitants having given a dinner to Governor Eyre on his landing . . . .[2]

Initially the impact of Darwin, at least within the small circle of anthropologists, was a liberal one. The first effect was to revive the debates between monogenists and polygenists. Briefly the former postulated a common ancestor for all human races and the latter considered the races to be separate species — descendants of separately created ancestors. This was not only a debate about how certain anthropological evidence was to be interpreted. It had political overtones. One of the chief protagonists of polygenism, James Hunt, President of the Anthropological Society of London, was also firmly against the abolition of negro slavery. Hunt had a strong obsessive dislike of the negro which he frequently expressed. The popularity of polygenist anthropology among intellectuals in the slave-owning states of America testifies to the political dimension which this sort of anthropology rapidly acquired in certain circles.[3]

The implication of Darwinism was that not only the human species but species in general were probably descended from a common ancestor. It also suggested that species were not so rigidly separated and absolutely distinct as had been thought. It gave succour to those who insisted that the religious or political unity of mankind could be backed by scientific evidence of a common physical origin. C. S. Wake insisted that a consistent advocate of the Darwinian theory,

grants to all men, if not a unity, yet an equality of origin, and he explains the mental peculiarities of different races by the influence of external conditions of life which have acted on the physical organism, favourably or unfavourably, and thus hindered or assisted the development of the mental faculties.[4]

However Darwin had also resurrected a graduated chain of development from ape to man and it was possible to combine this with theories of inherent difference and inferiority. It could be argued, for example, that, regardless of the unity of man, Darwinism implied 'primitive' peoples were early and inferior historical forms. Darwin's *Descent of Man* was an attempt to find a graduated series of links — mental, social and moral — as evidence for evolution. To this exigency, a belief in human equality, to which other areas of Darwin's life and work testifies, was sacrificed.

This represented a shift in emphasis. J. S. Mill, for example, had considered the theory of psychological graduation as 'one of the chief hindrances to the rational treatment of great social questions'.[5] E. B. Tylor also considered that,

It appears both possible and desirable to eliminate considerations of hereditary varieties or races of man and to treat mankind as homogeneous in nature although placed in different grades of civilisation.[6]

But there was only a thin dividing line between the subtle ethnocentrism represented by Tylor's theory of religious and intellectual evolution and the more strident insistence on mental and physical graduations between the races. Many anthropologists were drawn irresistibly to the conclusion that the graduated evolution which Darwin described in the *Descent* applied to intellectual and moral qualities and this was the best explanation of the cultural differences recorded in Tylor's theory of 'primitive' religion.

II

Characterisations of race and class related to a general typology of social hierarchy. It had these features. First of all the hierarchy was generated outside of society by natural rather than by social laws. A difference arose as to whether these social laws were generated by God or by natural scientific processes but, in both cases, the production of social categories had some basis transcending or outside society itself. The conviction of many social Darwinist thinkers that racial and social inequality was the product of natural selection is very little different from Paley's conviction that the pattern of both the natural and social world was generated by divine reason operating to distribute individual or classes of animals in their places.

In 1866 Wallace had warned Darwin about the phrase 'Natural Selection'. He believed it would be misunderstood and that the terms 'Nature' and 'Selection' would tend to imply a directing intelligence in evolution. But to warn against a form of words is to misunderstand a fundamental intellectual and social process. The belief in Direction and Intelligence generating history would have been introduced regardless. Behind this were social pressures which required that the social order be pictured as a pattern of rational categories — in other words that society was as it was because this was the best or most reasonable way it could be. A second major characteristic was to describe society and social and racial places in it in terms of individual attributes. What social Darwinists saw in natural selection was not dispelled by Wallace's alternative title 'survival of the fittest'. The latter term neatly coincided with another of their basic premises — that social position and social action is a function of individual faculty.

To give an example, the anthropologist Haddon described Empire in terms of individual qualities.

the statement that the most efficient peoples must ultimately prevail may be accepted as correct. The racial, economic, social and political history of South Africa affords us a striking example of this process in the mutual relations of Bushman, Hottentot, Bantu, Boer and Briton.[7]

Imperialism was often reduced in this way to individual faculty generalised to become the faculty of a group. But so too was the economy and politics. All seemed to obey the law of psychological characteristic and to express initiative, thrift, 'drive', ambition, intelligence or the lack of these qualities.

The third characteristic was the use of familial imagery to indicate inferiority. A tension existed between the language employed in mid-nineteenth century culture which implied legal, political and economic equality and the actual facts of subordination and inequality in society. Theoretically at least the precepts of political economy suggested a potential equality of access to capital and goods. The law distributed — in theory at least — equal obligations and rights and equality of punishment. Eventually the political system was modified to take greater account of political equality. But a contradiction existed between the theory of legal, political and economic equality and the existence of hierarchy. When it came to describing subordination the Victorians took much of their imagery from an area where subordination was legitimised — that of the family. Thus they intertwined the language of political and legal equality with that of the family to find a means of reconciling the fact of subordination with the precepts of a system which theoretically rejected it. Thus they talked in terms of dependence, of development, of benevolent and paternal supervision and of the 'child' or the childlike qualities of the 'primitive' peoples. This allowed a theory of the inherent dependence and inferiority of certain groups to be combined with the assumption that an underlying identity of common interest existed in spite of it. Even the British socialist used language of the family to describe the future socialist order or to justify the progress of social welfare. They argued for the recreation of emotional and moral ties between peoples in imitation of the ideal of the family. Similarly the colonial reformer attributed a paternal role to colonial government and used the idea of development towards 'maturity' as an image of the political emancipation of colonial peoples. Even the term 'dependency' came to describe a system of political subordination.

The fourth major determining factor of theories of 'inferiority' was the connection between social and racial inequality. The same typologies were used in both cases. Later in the nineteenth century, these 'racial' typologies became fixed more strongly upon race in the sense in which we now use it — upon relations of colour. But many of the ways of picturing both the black and the working class were basically drawn from one interconnected set of ideas. Galton put this connection perfectly when he claimed that,

Besides these three points of difference — endurance of steady labour, tameness of disposition and prolonged development— I know of none that very markedly distinguishes the nature of the lower classes of civilised man from that of barbarians.[8]

In fact typifications of race and class were interchangeable in much of the writing on the subject. Many of the racial categories of the late eighteenth and early nineteenth century were, in fact, generated initially as comments on the internal political development of certain countries — such as France leading up to, during and after the French Revolution. These were used primarily as descriptions of social groups within a nation and were intended to legitimise political and social divisions within it. Thus to the racial theorist and defender of aristocratic privilege, Gobineau, black, white and yellow referred not only to the great racial divisions of the world but also to the classes within France itself all of whom were believed to exhibit in their social and political behaviour some of the 'racial' characteristics of these three types. In fact it would be argued that it was from his categorisation of the classes within France that he deduced his racial picture of the world at large.[9]

In England racial categorisation showed a similar convergence between notions later to become associated almost exclusively with 'colonial and primitive' peoples and descriptions of the working class. That these characterisations came to apply largely to the colonial and the 'primitive' requires an explanation. But even in the late nineteenth century the network of ideas about 'inferiority' still showed considerable plasticity in the groups to which they were applied.

Nearly all the versions of social Darwinism we have touched on fitted the conception of a special British racial destiny. In addition most of those which expressed the superiority of the white race also expressed the superiority of the middle and upper classes. In 1879, for example, Alfred Marshall used the idea of the differential birth rate with both the poor and the 'inferior' races in mind.

There can be no doubt that this extension of the English race has been a benefit to the world. A check to the growth of the population would do great harm if it affected only the more intelligent races and particularly the more intelligent classes of these races. There does indeed appear some danger of this evil. For instance, if the lower classes of Englishmen multiply more rapidly than those which are morally and physically superior, not only will the population of England deteriorate, but also that part of the population of America and Australia which descends from Englishmen will be less intelligent than it otherwise would be. Again if Englishmen multiply less rapidly than the Chinese, this spiritless race will overrun portions of the earth that otherwise would have been peopled by English vigour.[10]

Alfred Marshall saw economic history as a product of certain psychological characteristics. The rudimentary character of economic organisation in early society was due, in his opinion, to the fact that 'Children and nations in an early state of civilisation are almost incapable of realising a distant advantage.'[10] Consequently they did not save and they worked only when forced to by necessity. But this was also true of the poor of his own society and in fact accounted, in some measure, for their poverty. 'Again the poor even in highly civilised countries are careless as to the future.'[10] They were 'too intent on satisfying their immediate needs to have time or inclination for forethought'.[10] In contrast the capitalist was distinguished by psychological characteristics such as ambition, thrift and foresight.

The comments of C. Lloyd Morgan, the animal psychologist, show clearly how biological and social beliefs were intertwined. Morgan wrote that,

In every group of animals there should be lowly forms which have remained stationary while their fellows were becoming the winners in life's race; just as we see the poor, the weak, the unenlightened, and the unsuccessful living on beside the rich, the strong and the highly cultivated and the successful. And these lowly forms should retain features which are embryonic in the more highly developed forms; just as the less favoured individuals among us are apt to be childish and undeveloped.[11]

Morgan was in fact describing an evolutionary zoological order but the social and biological orders were, in his mind, analogous to each other.

This cross-fertilisation of biological and social metaphors was important. It made the biologist the natural guide of the social sciences — the man or woman who could make conceptions of the social order 'rational' and 'scientific'. A. C. Haddon, who helped to found anthropology as an academic discipline at Cambridge and London, moved easily between the worlds of zoology — his original academic interest — and anthropology. In both he was an observer and collector. In both he applied similar notions of order and hierarchy. In his study of the shellfish 'Actinae' in the 1880s he applied von Baer's theory of recapitulation to establish the early form of the species and, in a similar fashion, he sought, 'in the youngest man the story of the oldest man'.[12] Man's cultural life was placed on a line of development parallel to that of the zoological order and in both the key to discovering the order was the investigation of the development of the young of the species. Haddon contended therefore, 'there is not only a parallelism to some extent in physical features between children and certain savages but there is in children a persistence of savage psychological habit, and in the singing games of children a persistence of savage and barbaric practice'.[12] The difference between the child of the white race and that of the black was that the former was considered able to progress towards adulthood whilst the latter was able only to raise himself a few steps in the scale of civilisation. He was one of those early forms which remained stationary.[12]

Biology helped therefore create the kind of moral universe in which nature reflected society and vice versa. This idea of a parallel between the two had been popular before Darwin. It was expressed most succinctly in Paley's natural theology. The immediate impact of Darwin was, in fact, to shatter the universe by undermining the notion of stability and lack of change in nature. Darwin's *Origin*, by emphasising the common descent of the races, also provided ammunition against the notion of separate creation of the species which, for example, Hunt of the Anthropological Society of London used to justify the negro slavery of the southern states of America. But in the *Descent* Darwin lent his authority to the notion of the separation of the races along a graduated evolutionary chain of development. The assumption was that the 'primitive' races represented early stages in the evolution of man. He also attempted to prove that psychological faculties such as intelligence and moral sense were also part of a graduated evolution. This implied that some races 'lower' on the evolutionary scale had also 'inferior' development of these faculties. Both of these concepts had little to do with Darwin's own politics or social conceptions. They had a great deal to do with his marshalling of evidence of human evolution in defence of his main hypothesis of natural selection. Nonetheless they did help recreate a post-Darwinian social and natural hierarchy which had a politically tendentious tone.

Wallace disputed the notion of a graduated moral and intellectual development. His version of evolution laid emphasis on the importance of reason and its emergence in human evolution. But he assumed that this involved a break with natural selection and that thereafter there would not be much progressive evolution in man. Nonetheless Wallace's early writings also show a belief in racial inferiority. By the end of his life, however, Wallace had modified those views. As social Darwinism emerged it became increasingly antipathetic to Wallace's socialism. Wallace consequently reconsidered the position of 'primitive' man and by 1913 he believed that,

Many other illustrations of both intelligence and morality are met with among savage races in all parts of the world; and these, taken as a whole, show a substantial identity of human character, both moral and emotional, with no marked superiority in any race or country. In intellect, where the greatest advance is supposed to have occurred, this may be wholly due to the cumulative effect of successive acquisitions of knowledge handed down from age to age.[13]

The conventional social and biological view was, however, the opposite of Wallace's. It recast society from the apex of social and economic privilege. Natural selection was expected to confirm this even to the extent that Wallace found when a correspondent contested his views; it was 'rather amusing to be told now that I do not know what natural selection is, nor what it implies'.[13]

Race was an idea easily adapted to Marshall's view that the character of societies was determined by the attributes of individuals and that the social hierarchy reflected differences in psychological aptitude. The notions of social hierarchy and race merged, firstly, in the assumption that the 'inferior' races were always destined to occupy a lowly social position and, secondly, in the belief that the existing class system in Britain was also a racial system. Attempts were made to depict the classes in Britain as physically as well as psychologically distinct.

Taking the population of England in a general sense the upper classes and the country folk seem, on the whole, to be fairer and taller than the industrial sections of the population; a disposition which may indicate a natural drift of the northern race towards modes of life giving openings for their directive and organising powers and to their love of a free life in the open air.[14]

This view of Britain's racial composition was often combined with a pessimism about the effects of urbanisation and industrialisation.

It seems probably, then, that these modern tendencies of our civilisation favour selectively the racial elements of Southern origin, the elements that, as far as we can ascertain, have been least productive of ability and genius in England. If this be the case, bearing in mind the characteristics of the two races, the British nation . . . may find themselves becoming darker, shorter, less able to take and keep an initiative, less steadfast and persistent and possibly more emotional.[14]

This crossing of racial and class stereotypes found a perfect expression in the Aliens Act of 1905. Not rigid enough for those for whom race was all important, its rigidity was, in fact, modified by the value it placed on class. It therefore excluded from examination by officials, aliens in first-class passage, allowed the second-class passenger to be excused from examination under certain circumstances and permitted the entry of those carrying five pounds or above.[15]

III

What effect did the ideologies of race have upon actual social practice? They acted as a means of organising behaviour in conformity with certain kinds of social relation. This was very true of anthropology. Westermarck, for example, tried to advance the discipline of anthropology by offering it as part of colonial government.

I am convinced that in our dealings with non-European races some sociological knowledge, well applied, would generally be a more satisfactory weapon than gunpowder. It would be more humane and cheaper too.[16]

In Westermarck's opinion the application of anthropology would have prevented the Indian Mutiny and also the 'recent troubles in Morocco'.[16] W. H. R. Rivers too believed that a bond should exist between colonial administration and anthropology.

It is a widespread popular idea that the chief tasks of the anthropologist are the measurements of heads and the collection of curious or beautiful objects for museums . . . . The main interest of the anthropologist today . . . lies in the regions concerned with the structure and organisations of human society . . . whatever may have been the preoccupations of the anthropologist in the past, his chief interest today is in just those regions of human activity with which the art of Government is daily and intimately concerned.[17]

It is true that there was an element of special pleading in these requests. A great deal of the finance for anthropological education rested on convincing the Colonial Office of the 'relevance' of it. But it is also true that the development of anthropology was closely bound up with that of Empire. This may have produced perfectly unexceptionable texts of knowledge. However, it is also true that it meant it was unlikely that anthropology would escape certain highly ethnocentric and political *parti pris* assumptions.

This is also the case with the emergence of academic sociology. The creation of 'scientific' sociology was, in fact, the recasting of certain ideological notions in new 'technical' forms — including, of course, 'biologising' it. Francis Galton, for example, drew up a questionnaire for colonial teachers which showed how indistinguishable fact and value were in much of the work done at this time. Entitled 'Ethnological Intelligence of Different Races' it included the following questions:

Question 10
Children of many races are fully as quick and even more precocious than European children, but they mostly cease to make progress after the season of manhood. Their character changes for the worst at the same time. State if this has been observed in the present instance.

Question 20
Children of savages, who have been reared in missionary families have been known to throw off their clothes and quit the house in a momentary rage and to go back off to their people among whom they were afterwards found in contented barbarism. State authentic instances of this, if you know any, with full particulars.[18]

Similarly the first question paper on ethnology set for the University of London sociology examination was based on a comparison between the savage and the child. It asked students: 'What are the principles that should guide the investigation of mental process in races of low culture. Consider how far these correspond with those appropriate to the investigation of child psychology.'[19]

The importance of the development of sociology as a discipline was that it was through these means that ideology of this kind found an institutional as well as a 'scientific' expression. The first sociology syllabi were in fact directed to those whose work brought them into contact both with 'races of a low culture' and with the metropolitan poor. These were described in a first syllabus as,

Borough Councillors, Poor Law Guardians, Members of Committees of Philanthropic Institutions and Societies, District Visitors, Trade Union Officials, Scripture Readers, Workers in Settlements, Rent Collectors, Workshop and Factory Inspectors, Friendly Society Workers, Officers of Benevolent Societies and in addition, in so far as Ethnology is concerned, Civil Servants destined for the Tropical portions of the Empire and Missionaries.[19]

The dialogue about race and class which social Darwinism had become was strongly centred on the question of authority. It conveyed, firstly, the inadequacy of those who deviated from a certain conception of cultural and intellectual development. Secondly, by characterising the faculties of those who deviated as, in some way, childish and lacking in restraint it conferred the right of guidance over them onto those individuals who had acquired the values of evolutionary progress or could acquire them by a study of sociology. Sociology was directed strongly at the stratum of society who stood in a crucial relationship to the 'less advanced'. Under the guise of scientific study it transmitted these values to them and through them to the recipients of charity, education and administration. The society which gave rise to these notions of hierarchy found a means of reinforcing them via an apparently 'scientific' training.

It was consequently obsessed with the question of mental and moral difference. It asked the student to identify 'the chief mental criteria of value in differentiating race'[19] and 'for evidence . . . of correlation between intellectual development and social structure'.[19] Also it introduced students to a question upon which the possibility of evolutionary development among 'savage' peoples was considered to depend — 'How far does the history of primitive peoples give evidence of a real increase of mental capacity.'[19]

Even more important was the role given to moral evolution. This evolutionism was at the furthest remove from the biological determinism of eugenics. It was nonetheless highly ethnocentric. It consisted of a series of moral and intellectual tests given to other cultures and races. 'Give facts illustrating the savage's idea of personality, and consider how far this differs from *the fully developed concept*'[19] ran one such question. Another asked 'How far does the treatment of strangers by any social group indicate the *moral level* reached by that group?'[19] In a similar vein students were given questions asking them to 'Illustrate from the history of punishment for crime the *development of the recognition of personal responsibility*.'[19]

It was difficult to negotiate questions of this sort without conceding the case that some social groups were 'backward'. This was true even though the majority of early sociologists rejected the idea of the crude 'struggle for existence'. This did not necessarily mean that they abandoned the belief in European racial superiority. For example, Leslie Stephen battled with the contradiction between the evolution of morality and the struggle for existence'. But he came to the conclusion that moral evolution did not hinder Imperial conquest. On the contrary the dominant race 'holds its own, not merely by brute force but by justice, humanity, and intelligence, while, it may be added, the possession of such qualities does not weaken the brute force where such a quality is still required'.[20]

Stephen's version of evolution laid strong emphasis upon cultural superiority and it also held out the notion of a convergent evolution between peoples. Culture and ideas could be transmitted between societies. Benjamin Kidd also reaffirmed the importance of cultural evolution. It was 'the nature of social heredity which creates a ruling people. It is what it lacks in its social heredity that relegates a people to the position of an inferior race'.[21] However late nineteenth-century social theorists began more and more to treat social difference as a product of psychological difference and therefore social 'inferiority' as the consequence of psychological 'inferiority'.

This meant that however 'liberal' the theory of cultural evolution might appear, some peoples were considered to be excluded by their mental backwardness from evolutionary development.

Certain characteristics of British intellectual and social life acted to keep sentiments of this kind within bounds. The doctrine of evolution could for example support a belief in limited political progress for the 'backward' races. This depended upon the capacity of the savage for intellectual and moral development. If this could be proved then ultimately his emancipation was possible. But this was an even more important question in the case of the 'savage' within Britain. Most of Hobhouse's concern about eugenics and heredity stemmed from the problem that would be created within British society if it could be shown that the inadequacy of the poor arose from some deep-rooted and ineradicable hereditary flaw. All those who believed in drawing the classes together and in the importance of education and moral training as a means to this end found the prospect of an hereditary gap between the classes frightening in its implications. This would transform the social missionary into an agent of supervision and control rather than education and enlightenment. It made the antagonism between classes seem ineradicable by giving it the sanction of physical and hereditary difference.

Thus Hobhouse's interest in the 'bearing of different theories of heredity on questions of social progress'[22] concerned his own society rather than the relations between that society and other cultures. Most liberals and socialists even whilst they held out hope of colonial emancipation believed the character of evolution meant that, at the very least, some form of benevolent tutelage must continue to be exercised over the 'inferior' races.

Ramsay MacDonald, for example, argued in *Labour and the Empire* in 1907 that,

The white nations which exploit the Tropics economically assume responsibility for the natives, and how to fulful that responsibility is the kernel of the problem of dependency government. This responsibility, however, may be regarded from a worthier point of view than as a consequent of economic exploitation. A community may well claim that it has a duty imposed upon it to spread the blessings of its civilisation over the earth. *Morality has a universal sway*, and by reason of its *imperium* the more developed nations are brought into a position of something like guardian and teacher of the less developed nations.[23]

Notions of this sort neither insisted on racial equality nor on the destruction of Britain's colonial links. They merely suggested that colonialism was a transitional stage and argued that Britain should act more humanely in her colonies.

Real opposition tended to come from a section of liberals who argued that the politics of race and Empire had disastrous effects upon British society. They argued more often from this position than from the idea of any *à priori* equality between the races which frequently they denied. This comes out clearly in the debates in the 1860s on the emancipation of the negroes in the southern states of America. J. E. Cairnes, for example, argued that the liberal state depended upon the belief in political and legal equality. If this was breached in the case of the negro — whose capacity for self government and political advance Cairnes, unusually for that time, defended — then the principles of liberalism themselves came under attack. But perhaps more importantly the institution of slavery produced a class of rulers hostile to any extension of freedom. These had the attributes of a new feudal class and they introduced into metropolitan politics the values of feudalism.[24]

Leslie Stephen and T. H. Huxley expressed this view clearly. Although — unlike Cairnes — both believed in the inherent inferiority of the black man they regarded it as best for their own society if he were left alone. According to Stephen, the black man could be trusted with a limited amount of self government.

The white man, however merciful he becomes, may gradually extend over such parts of the country (Africa) as are suitable to him, and the black man will hold the rest and acquire such arts and civilisation as he is capable of appropriating. The absence of cruelty would not alter the fact that the fittest race would extend . . . but it may ensure that whatever is good in the negro may have a chance of development in his own sphere.[25]

T. H. Huxley also argued,

It may be quite true that some negroes are better than some white men; but no rational man cognizant of the facts, believes that the average negro is the equal still less the superior of the average white man. And if this be true, it is simply incredible that, when all his disabilities are removed and our prognathous relative has a fair field and no favour as well as no oppressor, he will be able to compete successfully with his bigger brained and smaller pawed rival in a contest which is to be carried on by thoughts and not by bites. The highest places in the hierarchy of civilisation will assuredly not be within reach though it is by no means necessary that they should be restricted to the lowest.[26]

Nonetheless Stephen conceded to the black man 'his own sphere' and Huxley argued strongly for the abolition of slavery and the possibility of some political and social improvement for the negro. Mid-nineteenth century liberals associated anti-abolitionism with the strengthening of the aristocratic influence in British political life. They also saw the arguments about racial equality in the context of the movement towards the extension of the suffrage which was taking place in the 1860s. They are not unaware of the interconnection between theories of racial inferiority and those of the hereditary inferiority of other social groups. To concede one half of the equation — negro incapacity for self government — could be taken as conceding political inequality in England. This was in effect to abandon the struggle towards the liberalisation of the British state.

As Britain's Imperial commitments expanded the implications of Empire upon her internal politics led to the further alienation of a number of liberals like Wallas, Hobhouse and Hobson from colonialism. Stephen and Huxley had written at a period when Britain's colonial commitments were much smaller. Wallas, however, saw the effect of colonialism and racial politics in an age of democratic politics. Liberals had hoped that democracy would strengthen the liberal elements in British society. However, Imperialism in an age of admittedly limited democracy had become the kernel from which some liberals believed a new tyranny might very well emerge in British political life — one with a popular mandate. Wallas listed the paraphernalia of racial politics. Even overtly 'liberal' issues like the 'Chinese Slavery' question in the election of 1906 cloaked the use of racial images and symbols. Behind this lay an appeal to racial feeling, a mobilisation of the crown against the individual and the exploitation of suggestibility. The politics of racial or 'herd' feeling signalled in his mind and that of Hobson the end of any chance of a political life based on decency and rationality.

This implied a drastic revision of the theory of moral evolutionism. Apparently the gains made by social evolution were more hardly won and precarious than earlier liberals had realised. In these circumstances the discussion of whether the black man was capable of self government was largely irrelevant. Hobhouse thought it possible that 'The experience of Cape Colony tends to give the affirmative view. American experience of the negro gives, I take it, a more doubtful answer. A specious extension of the white man's rights to the black may be the best way of ruining the black.'[27] But the proper solution was, Hobhouse believed, not to get involved even as educator. 'Until the white man has fully learnt to rule his own life the best of all things he can do with the dark man is to do nothing with him.'[28] This was an admission that British liberal democracy hung on too slender a thread to allow into it the politics of Imperialism.

But even this current of opinion attempted to express itself in biological metaphors. Wallas's revulsion against much contemporary politics was framed in terms of Darwinian instinct. Hobhouse wanted a scientific backing to a theory of evolutionary moral development. Even Hobson who went further than most in removing theories of Imperialism from 'biology' resorted to the language of social Darwinism in his discussion of metropolitan politics. Partly this was because, to some, social Darwinism had become the language of social unity. Liberals expressed their desire for the coming together of classes in the metaphors of organicism and their hope of social progress in terms of moral and intellectual evolution. In rejecting a 'Darwinian' theory of colonial expansion in favour of an economic one, Hobson could be said to be indicating his unwillingness to see Imperialism as part of the scheme of evolutionary progress and his determination to treat it as an unpleasant residue of evolutionary backwardness in British society.

Hobson's *Imperialism* (1902) was written in the context of defences of colonial expansion which insisted that they were an expression of biological laws. According to Charles Harvey in a book entitled *The Biology of British Politics*, (1904),

The history of civilisation is a record of the continual substitution of combination for competition . . . the formation by combination of new collective forms.[28]

Imperial expansion was, according to this view, merely the logical extension of this natural process in the social sphere. Whereas Hobson saw it as the result of the inadequacies of social organisation and as fuelled by atavism this interpretation of Imperialism made it part of cultural evolution. It treated it as an extension of rationality to the international sphere. The notion that Empire was a natural law of the amalgamation of units even influenced socialist theories of colonialism — especially in the British context. This theory conceded prestige to the state and enhanced its role. As Bernard Semmel has shown, it led to demands for the active intervention of the state in areas of social welfare to improve the strength of the people and hence Imperial efficiency. This proved an attractive idea to some socialists. As Imperialism made increasing demands on national and patriotic feeling, it seemed to propagate the kind of cross-class social solidarity which socialists like MacDonald believed to be the moral basis of socialism and the end to which evolution was tending. Finally Imperial expansion was seen as the movement towards rationalisation of state systems of which socialism was also an expression. Other socialists of a different persuasion argued that this amalgamation of states in Empires permitted the transcendence of purely national feeling — a view which from the perspective of history looks, at the very least, naive.

IV

Social Darwinism became a means of explaining the differences between races and between classes. It showed great flexibility in the attitudes it could express to these questions. It could be both pro- and anti-Imperialism; in favour or against the extension of social reform; for the equality or inequality of the classes. This flexibility reflected the real contradictions and confusions about social policy and colonialism which existed. However there were limits on the flexibility. Intellectual life in Britain was not an equal representation of all views. Overwhelmingly social Darwinism was a justification for existing social relations and a vehicle for a belief in the inequality of race and class.

So long as subordination existed so did the language of subordination. It is also true that certain social developments acted to modify it. Historical events — such as for example the French Revolution — replaced or attempted to replace one political vocabulary by another. It attacked the language which embodied a theory of inferiority as well as some of the institutions which created and maintained that inferiority. Similarly the rise of political democracy in Britain modified the application of the language of racial inferiority to the working class. This did not mean it disappeared. As we have seen the Whethams in the early twentieth century were still trying to find a physical racial basis to class which, in their opinion, paralleled the relations between black and white. In addition much of the language of inferiority was generated in relation to economics rather than politics and this limited the effect of the growth of political democracy. So that, for example, the idea that unemployment was the product of the moral inferiority of the unemployed remained a persistent theme in writings on this subject well into the twentieth century even when the notion of 'citizenship' was being extended to more and more groups in society. The language of 'primitiveness' was still employed to explain the economic condition of the poor. To reformulate Galton's view of the lower classes of civilised nations, unemployment was attributed to the lack of 'endurance of steady labour, tameness of disposition, and prolonged development'.

But it is also true that the gradual admission of the working classes into the system of political democracy acted as a check upon their racial designation. It also had the concomitant effect of refocussing this language more exclusively upon black and white. The idea of 'backwardness' came to have the appearance of having been framed specifically for the 'native' — a belief which tended to underline the twentieth-century notion that racial language was developed in order specifically to describe black and white. In fact its origins lay in the typologies by which all sorts of 'inferiority' and subordination were explained and justified.

Theories of racial inferiority were not products of 'mistaken' scientific notions generated in the nineteenth century and corrected in the twentieth, whose persistence, therefore, can only be the product of ignorance or irrationality. Ignorance and irrationality accompanied racialist theories but there were underlying social not personal causes for their survival. The catalogue of mistaken scientific ideas which gave apparent legitimacy to racialist theory is a long one. The fact that they survived despite the demonstration of the intellectual emptiness requires a social not intellectual explanation. Darwinism dispelled a number of racialist ideologies. Later on, Darwinian language was reincorporated into other nineteenth-century theories of racial inferiority. This illustrates the power of social forces to reassert themselves over language and ideology. Darwinism was born into this world, it did not create it. The ability of many twentieth-century intellectuals to detach themselves and areas of science from the conception of race dominant in preceding generations lay not only in their discovery of 'mistakes' but because of the impact of social and political developments.

Nor was racialism purely a product of a conceptual system exercising power over social life and ordering its 'practices' and relationships. Conceptual systems which generate power over life have a relationship to some objective social force. That is why the discourse of race we have today is not that of the aristocratic theorists of the late eighteenth and early nineteenth centuries. The growth of nationalism and political liberalism made these versions redundant historical curiosities unless, as happened in Gobineau's case, their precepts were substantially transformed to meet new social conditions.

IX THE LEGACY OF SOCIAL DARWINISM

WHAT became of social Darwinism? One theory asserts that at the turn of the century a reaction developed against evolutionist and stage theories of human society. This was represented by the anthropologist Malinowski's criticisms of the anthropology preceding his work. The main shortcomings he attacked were,

In my opinion, they always centre round the question, whether, in constructing an evolutionary stage system, or in tracing the diffusion of this or that cultural phenomenon the scholar has devoted sufficient attention to the full and clear analysis of the cultural reality with which he deals.[1]

In other words, subsequent social theory discarded the obsession with evolution and instead concentrated upon the actual social processes in society. Because of this, it is argued, social Darwinism was largely abandoned.

In fact, this new concentration upon the social system was largely built around organicist theory which had played such an important part in the social philosophy of the 1870s and 1880s. This organicism was not necessarily Darwinian but it was certainly biological. It consisted, in fact, of treating the social system as though it was some supra individual from whose 'needs' for survival and efficiency certain social institutions and arrangements emerged. This tradition, though it was subsequently refined and elaborated, continued to play an important part in social theory.

In *A Scientific Theory of Culture* (1944) Malinowski listed these organic needs of society. They were material production, a set of rules by which social behaviour was controlled, a system of education by which these rules might be instilled in the individual and political authority which co-ordinated social life and activity. The *raison d'etre* of these institutions was the survival of society. Just as human 'faculty' could be measured in terms of its contribution to the evolutionary 'fitness' of the individual, so these cultural attributes could be judged by their contribution to the survival of a particular culture. In a critical essay on Malinowski, Talcott Parsons argued that the prerequisites of society as outlined by Malinowski were too wedded to basic biological interpretations of 'needs'.[2] In contrast, Parsons argued that his own set of 'functional imperatives' included cultural as well as biological 'needs' of the social organism. However, in both cases what was being attributed to society were 'needs' and the language and concepts in which these were expressed were largely derived from a form of biological individualism projected onto a larger social scale.

The difference between this organicism and that of the late nineteenth and early twentieth century lay in the fact that the organicism of the functionalists represented by Parsons and Malinowski was less philosophically orientated. In this respect it *was* more concerned with actual social processes. Though Stephen arrived at a theory in which the individual, social institutions and society as a whole were treated as an interconnected system, he did not go very far in linking this with actual description of particular institutions and societies. Nor did he look at each of these areas individually. However, Stephen had justice on his side in considering, as F. W. Maitland reported, that *The Science of Ethics* was a considerable achievement in formulating a broad general social theory.

Malinowski considered functionalism implied a shift from evolutionism but not from biology. Even then he considered that, 'Evolutionism is, at present, rather unfashionable. Nevertheless its main assumptions are not only valid, but also they are indispensable to the field worker as well as to the student of theory.'[3] This view was justified for when Malinowski returned to a consideration of the character of evolution, he took up again the concepts of social Darwinism,

In biological evolution the concept of the survival of the fittest and the struggle for existence still retains its fundamental importance, in spite of certain corrections which were inflicted upon it by Darwin's followers. Prince Peter Kropotkin was quite right in pointing out that mutual aid between individuals of a cooperative community is the dominant concept, while the struggle between the individuals for survival can not be applied to human societies as a whole. We could not intelligently and with any chance of documentary evidence apply the concept of struggle for existence to primitive communities, certainly not in the sense of assuming a perpetual state of warfare, of extermination of weaker groups and the expansion of stronger ones at the expense of those defeated or destroyed. We can, however, apply the concept of survival value to cultures. This probably would not be coupled here with any concept of struggle, but rather with that of competition, within cultures and between cultures. We could affirm that the failure within any culture as regards instrumental efficiency, artifacts, cooperation or symbolic accuracy would inevitably lead to the gradual extinction of the whole cultural apparatus.[3]

This view was not very much different from the notion of cultural competition present in Stephen and Bagehot. Malinowski's adherence to it would justify Parson's opinion of him that he failed to break sufficiently with a certain nineteenth-century tradition of social thought.

However, even the most highly 'functionalist' social theorists of the twentieth century formulated their theory of change in a social Darwinist manner. Kingsley Davies in 1964, for example, puzzling with the origin of value systems considered that they,

resulted from the process of natural selection on a societal basis. In the struggle against nature and in the struggle between one human society and another, only those groups survived and perpetuated their culture which developed and held in common among their members a set of ultimate ends. The important thing was not so much the particular content of the ends but rather the fact of having ends in common. Viewed in this light the possession of common ends must be as old as human society itself.[4]

This notion was similar to Darwin's exegesis in the *Descent* of the importance of the social instinct in evolution and that of many writers who followed Darwin on the importance of moral codes in upholding social solidarity.

Twentieth-century functionalists did not provide much more than an elaboration of these ideas. Their organicism, in fact, forced them to think of social evolution as a form of individual 'adaption'. Society tended to be treated as though it was an individual among a series of other individuals in a struggle for existence. Cultural institutions became analogous to individual faculties and their 'superiority or inferiority' a measure of the extent of evolutionary development. Thus H. M. Johnson talked of a society responding to a social or non-social environment.[5] Parsons considered that social adaption led to increasing complexity and differentiation in society — a view strongly reminiscent of Spencer's theory of evolutionary development, and, like Spencer, he tended to treat increased complexity as an indication of a 'higher' evolutionary place.

The functionalists also used the analogy of social variation and cultural selection. According to one theorist of cultural dynamics writing in 1947, progress was the result of invention. This covered not only technological development but also cultural.

Yet ideas are no less powerful than things in shaping the lives of men. It would be difficult to maintain that the 'inventors' . . . who devised the method of counting descent on one side of the family, or who later developed classificatory systems of relationship terminology had less influence on the course of human culture than had the inventor of the skin tent or of the outrigger canoe.[6]

Out of these inventions came cultural selection. But, as Talcott Parsons said, 'Unlike biological genes, cultural patterns are subject to "diffusion" '[7] so that the notion of cultural selection did not preclude, as Stephen and Bagehot both saw, a convergent evolution between societies.

When twentieth-century functionalists looked for the origin of social variation they often found it, like their nineteenth-century counterparts, in invention. According to Parsons,

From his distinct organic endowment and from his capacity for and ultimate dependence on generalised learning, man developed his unique ability to create and transmit *culture* . . . cultural innovations, especially definitions of what man's life *ought* to be, thus replace Darwinian variations in genetic constitution.[7]

Stephen would have fully agreed with this. He would also have argued that this made social evolution dependent upon the evolution of rationality. Similarly, twentieth-century theorists answered the question of where ideas came from in the same way. Cultural innovation was, in part at least, a result of the growth of rationality. This separated, in their view, 'primitive' from 'advanced' societies. Similarly so strongly was their concept of social complexity rooted in organicism that the other factor distinguishing social systems was held to be the degree of their 'differentiation'.

The twentieth century saw a shift in terminology. 'Primitive' became 'undeveloped' and the terms 'developed' and 'undeveloped' were used particularly in relation to the extent of industrial development. However, the explanation given for the development of industrialisation was still based on very similar premises.

The organisation of industrial capitalism represented the highest development of certain forms of rationality. Strangely enough this view was integrated with a proposition also found in Bagehot's and Stephen's work — that progress is the result of creative innovation. This notion became merged with Weber's idea of charisma. Bagehot's great man played a similar role as the instigator of new social 'variations'. Capitalism was therefore, for example in the work of Bendix, characterised by a combination of organisation and of the entrepreneurial ethic — the latter a form of 'great man' theory, or perhaps it would be better to call it a great thought or ambition theory.[8] The result was the idea that these mental energies could be released in particular sorts of societies — those where status was not ascribed but achieved and where innovation was rewarded.

Thus a characterisation occurred which was strongly based on Bagehotian notions of 'traditional' as opposed to 'modern' society. In the former there were 'institutional' obstacles to the development of individual initiative; in the latter initiative had an important and honoured place. Thus the solution to evolutionary 'backwardness' was in institutional change of the kind which nourished the development of a certain kind of rationality and of individual effort. To this was added the notion which Bagehot called 'atrophy' — the idea that societies which lacked the institutions presumed to be characteristic of 'advanced' nations had no history but remained immured in 'traditional' modes of political and economic life.

Like Bagehot and Stephen, these theorists believed in 'convergence'. In particular they believed there was no inner physical or mental difference between peoples which prevented the transmission of ideas. Therefore evolutionary progress was possible for all peoples. However, the functionalist theory of development was a highly ethnocentric one. As we have seen in the late nineteenth century it was quite possible to root notions of superiority in cultural difference. If societies were like species then the existence of different culture forms side by side might very well be expressed in this kind of language,

The surviving lower types, however, stand in a variety of different relations to the higher. Some occupy special 'niches' within which they live with limited scope, others stand in symbiotic relations to higher systems. They are not, by and large, major threats to the continued existence of the evolutionary higher systems. Thus though infectious diseases constitute a serious problem for man, bacteria are not likely to replace man as the dominant organic category and man is symbiotically dependent on many bacterial species.[9]

'Higher' and 'lower' cultural systems provided as much jus-tification for supervision and control as did inferior and superior mental abilities. Thus whilst representing a 'liberal' view of the future of the undeveloped countries, it represented one centred very strongly in an assumption of the superiority of industrial capitalism. And this superiority was not merely 'technical'. It also rested on the degree of rationality in a social system, the kind of economic initiative displayed in it and also in its institutions.

Functionalism was more of a social Darwinism than its protagonists realised. It related to one tradition at least of using Darwinian analogies. It also showed a similar high-handedness in relation to biological science as its close relation of the late nineteenth century had done. In Parsons' view variations became evolutionary 'universals' — an attempt to unite Weber with Darwin. However, 'universals' are varia-tions which the social system 'requires' to come into existence — therefore they are teleological demands made upon the organic form which it obligingly meets. To illustrate the 'biological' basis of this idea Parsons pointed to the brain and the eye which he considered were not necessarily universal but at least generally present in a great many evolutionary forms. Darwin considered the eye to have given him particular trou-ble. He believed, correctly, that it would frequently be quoted against him in defence of a preordained pattern of evolution.

He spent a great many hours pondering on how he could show that its existence did not prove direction and purpose in evolution. Darwin saw a kind of teleological argument in this and it is only fair to ask whether Parsons' social evolution is not a sociological version of Mivart's theory of evolution. Secondly, Parsons took it as axiomatic that development meant complexity of structure. Darwin did not. Even the general occurrence of complexity in later evolutionary forms did not prove its necessary existence any more than the evolution of the eye in many organisms proves that evolution exists to produce the eye.

Stephen, too, slipped into a teleology of evolutionary development. In a passage in the *Science of Ethics* Stephen put forward the view, not dissimilar from that of Parsons, that evolution was a process,

of discovering a maximum of efficiency, though the conditions are always varying slowly, and an absolute maximum is inconceivable. At every point of the process there is a certain determinate direction along which evolution must take place, the form which this represents is the typical form, any deviation from which is a defect.[10]

Evolutionary universals had certain features in common with this view. They increased social adaption and they were generally developed in all societies. Thus, in Parsons' view, a society which had not developed certain of them must, for that reason, be considered ill-adapted.

To a great extent, therefore, the functionalists repeated the mistakes as well as the good ideas of this tradition of social Darwinism. But there are other parallels. It could be argued that each strand of this tradition was in a similar social and political position. Both were involved in a defence of the intellectual in society. The former in the conditions of late nineteenth-century Britain, the latter for a while in those of American society during the Cold War. This makes the functionalists in certain respects liberal rather than conservative theorists.

However, the twentieth-century functionalists made organicism an even more conservative doctrine than Bagehot and Stephen. Bagehot believed that social order was, at least in early evolutionary history, the product of political and military despotism. He also believed that the bonds of social solidarity required, at times, rather artificial cultivation. He was not above attributing the social respect felt for one's 'betters' to the projection of an illusion designed to impress and flatter the less educated. Both Stephen and Bagehot emerged from a period in which liberals believed the notion of social variation justified further social and political reform. An excessive concern with moral solidarity at the expense of rational advance would have seemed highly politically suspect to them. For the functionalist, the reduction of the problem of social order to the maintenance of 'value systems' had the effect of dissipating politics altogether. It suggested that political dissatisfaction arose from individual moral failure, the failure of those institutions — family, school, church — whose function it was to properly instruct the individual, or from the inability of the citizen to integrate the variety of conflicting demands made by social institutions upon him or her. It is not surprising therefore that functionalism should have become associated with a period of dreary conformity. Although Stephen and Bagehot were highly selective about the areas of social non-conformity they would tolerate, this outcome would have surprised them.

II

There was, however, another trend of social Darwinism — that concerned with the question of 'fitness' and its distribution among the population — an individual rather than cultural theory of evolution. The eugenics movement survived the First World War and it survived the 'hiatus' in studies of heredity produced by Mendelism. By the 1920s Fisher had included a eugenic chapter in his *Genetical Theory of Natural Selection*. Interest in the evolutionary implications of the birth rate survived and a new set of eugenic propagandists arose. C. P. Blacker a representative of this new generation claimed, in 1926,

Among uncivilised peoples it is the biologically superior type which is most prolific. In advanced civilised countries which are democratic in social organisation the reverse obtains.[11]

Blacker listed three reasons for this; the advance in medicine, increase in humanitarianism and, lastly, the method of obtaining votes. This last characteristic, due to the advent of democracy, meant that the desire for social welfare was appeased regardless of the eugenic cost.

Eugenics continued to have influence. First of all on the birth control movement where the idea of controlling the population for eugenic reasons was taken seriously by Marie Stopes.[12] Blacker demanded birth control for several purposes — to balance the birth rate of the working classes against the professional middle classes, and to prevent war. This became, after the 1914–18 War, a popular theme of eugenic propaganda. According to the theory, war was the result of overpopulation. Thus spreading information on birth control appealed to a variety of groups. It appealed to the proponents of individual sexual freedom; it was part of the attempt to make family life more tolerable for the poor; it was seen as helpful in the movement to liberate women. Mixed up with these motives, as always, was an adulation of the middle classes. This mixture of radical and conservative ideas represented a clear continuity with pre-war eugenics.

The eugenics movement took a great interest in two measures of the 1930s; firstly the debate on family allowances and secondly the Committee on Sterilisation in 1934. In the latter case, ominous evidence of the watershed which the rise of Nazi eugenics was to provoke in the movement emerged. Parallel to the Committee, eugenic legislation was passed in Nazi Germany. The eugenics movement felt a mixture of apprehension and admiration at the progress of eugenics in Germany. There was a succession of German speakers on eugenics to the Society in the twenties and thirties. But the actual details of the eugenic measures which emerged after Hitler's rise to power were not unequivocally welcomed. Eugenists pointed to the USA as a place where strict laws controlled marriage but where a strong tradition of political freedom existed. But several of the proposals of the new German government emanated from a variety of racism unpopular with British eugenists. For example, in much British eugenic literature, the Jewish race was held up as an example of educational and professional achievement.

Nonetheless, the eugenists considered that,

Though some of the details might not meet with general approval in this country, the broad outlines, as so far sketched, of the German Bill will certainly command the assent of all experienced eugenists. It is therefore decidedly deplorable that the scientific tenor of the proposals should be entirely vitiated by the inclusion of 'foreign races' among potential sterilizees.[13]

British eugenists found it hard to countenance the anti-semitism of the German Reich and were worried at the inclusion of the category of 'undesirables' in the proposals about sterilisation. Though Blacker hoped that 'careful perusal of the Bill does not make it by any means certain that this interpretation is justified',[13] he too was not altogether happy with some aspects of Nazi eugenics.

The difficulties which the eugenics movement faced when the full character of Nazi eugenics became known after 1945 were considerable. It helped push the movement into temporary obscurity. But this was not the only reason for their loss of influence. The 1930s saw the emergence of a leftward trend among a small but influential section of intellectual opinion. This made consensus about the inferiority of the working class more difficult. For example, J. B. S. Haldane, who inherited Pearson's chair at London, conducted propaganda against eugenics in the thirties. The chair of social biology at the London School of Economics was occupied by L. Hogben who shared Haldane's opinion about the character of much eugenic propaganda.[14] Two strategic positions in the eugenic world were, therefore, occupied by anti-eugenists.

In addition to this, the eugenics movement had to take into account the rise of the Labour Party. Blacker was quite correct in identifying this as a limitation on the more radical eugenic proposals although perhaps he also overestimated its influence. But it did help cause a shift in the axis of British intellectual life, leaving eugenics at a point slightly further from the central debates on social policy than it had been in the pre-war period. The *Eugenics Review* identified the *Daily Herald* in the 1920s as an opponent. J. G. Crowther and Ritchie Calder worked in the twenties and thirties for the *Manchester Guardian* and *Herald* respectively. Both represented the labourist and left-wing tendency. They helped found science journalism and the thrust of their writing was strongly anti-eugenist. F. C. S. Schiller recognised this when smarting at the treatment the eugenics conference of 1926 had received from the *Daily Herald*; he argued that,

However, if the position is to be dragged into the mire of politics it may be remarked that the most dangerous as well as the most powerful of the opponents of eugenics are the conservative capitalists and militarists who mistakenly imagine that the country needs plenty of cheap labour and cannon fodder.[15]

Although eugenics did not dominate debate about social welfare to the extent it had done before the First World War, it still had influence. Paul Addison in his account of the formation of attitudes to social reform in the years leading up to the Labour Government of 1945 has referred to a 'mandarinate' composed of civil servants, intellectuals, welfare workers and 'experts' who helped to create what he refers to as a 'consensus' (before and during the Second World War) about the need for the implementation of a major programme of welfare reform.[16] Many histories of the provision of public welfare are inclined to see its extension either as the outcome of humanitarianism or, alternatively (and with some justice), as due to popular pressure. But there was in addition a group of intellectuals who had deserted the theory of laissez faire for belief in state intervention and who acted as a pressure group for the development and extension of social welfare.

Among these eugenics had considerable influence. Many intellectuals had reached maturity in the epoch in which social administration and eugenics were closely intertwined. Beveridge was one. By the 1940s other influences such as that of Keynes (himself a member of the Cambridge Eugenics Society before the First World War) had made a major impact on Beveridge. Nonetheless many 'mandarins' had developed 'technical' and 'scientific' principles of social administration in which the application of eugenic principles played a not inconsiderable part. Cyril Burt represents one such influence. Before the First World War he was psychologist to the London County Council. After the war he was increasingly involved in projects which brought him into contact with government departments — for example the Home Office, Ministry of Health and the Board of Education. He carried with him the belief that social administration could be based on biological and psychological categories; that the major class divisions in society were, largely, biological; and the different academic performance of individuals could be attributed almost wholly to heredity.[17] In many ways it was only when the fortunes of war and politics led to the expansion of social welfare that many 'technicians' of social engineering came into their own. They offered technical expertise and their training was often a eugenic one. Their opportunity to exercise this training may have been the result of a minor social revolution but the tradition from which it came was much older. The debate is still being carried on about how far its presuppositions affected the course of British social policy.

II

In anthropology as interest shifted to the social system, partly due to Malinowski's influence, less attention was paid to allotting places in a hierarchy of evolutionary development. However, attitudes which expressed a belief in the racial superiority of the white man still persisted as did the apparent need to find an anthropology which would confirm it. C. G. Seligmann exemplified this.[18] Seligmann became interested in anthropology during his work on the medical pathology of the native peoples of the Torres Straits in 1896. He became a lecturer in anthropology at the LSE and later Professor. In the 1930s he was still convinced that measurable physical and mental difference existed between the races — for example that a gulf existed between the European and the Australian aborigine which could not be accounted for by culture.

1. Both macroscopic and microscopic evidence point to there being qualitative differences in the brains of Australians and those of certain of the more culturally advanced races . . . . 2. Examination of Australian culture and intelligence tests, so far as they are applicable, furnish indications of intellectual difference between the less and more advanced of the primary subspecies (races) which may fairly be regarded as correlated with differences in brain structure and hence as 'racial'.[19]

Attitudes of this sort certainly persisted. However, certain factors tended to modify them. First, as far as anthropology was concerned, physical anthropology — craniology and anthropometry and the assumptions upon which it had been based — received a severe blow with the development of population genetics. This simply shifted attention away from the previous practices of racial classification. To speak of race gradually came to mean to speak of a genetic pool. To identify a genetic pool required quite different tools of analysis from those of traditional physical anthropology. This meant that physiological categorisation seemed less able to yield true racial differences.

But the attitude of anthropologists was also affected by the change in political climate in the 1930s. A. C. Haddon, who at the turn of the century had been convinced of the inferiority of 'savage' races, went a considerable way to a renunciation of his former beliefs in 1935 when he published *We Europeans* with Julian Huxley. This was an attempted demolition of many 'racial' myths using for that purpose some of the complicating factors which modern genetics had introduced into the problem of racial classification. But the book was also notable for another fact. Prominent among its concerns was the international situation rather than the colonial one. Much of its attack was directed at the ideas of European racial antagonism and its objective was firmly built around the desire to contradict some Nazi versions of European racial history.

To those of Haddon's generation (with some notable exceptions) the inferiority of 'colonial' peoples had been axiomatic. But they had never altogether approved of European versions delineating European history in terms of racial antagonisms. Haddon had published a geo-racial map of Europe but he had not considered that national conflicts corresponded to racial divisions. But by 1935 he was prepared to accept that the classification of Nordic, Aryan, Alpine etc. which had been applied to Europe had been rendered archaic by developments in science and noxious by its association with a certain kind of European politics.

Further, Seligmann, Haddon, and many of the founding fathers of British anthropology who were reared on Galton's anthropometric measurements had to admit that, in practice, they had proved of little worth. On the Torres expedition of 1896 few significant physical or perceptual differences had been elucidated by the use of measurements of sight, hearing, of the skull, of throwing and pulling. Myers and Seligmann persisted in their conviction of the value of anthropometry. So did McDougall. But Rivers turned to cultural diffusion as an explanation of social change and Haddon also eventually dropped anthropometry.

Racism, as the case of Nazi Germany showed, was cavalier in its attitude to certain scientific facts. But the belief in the inferiority of certain races appeared to have a 'survival value' which transcended the question of proof. To paraphrase Parsons, one could ask what aspect of cultural adaption racism represented. If one scientific system went, another would do.

The struggle to reassert some kind of measurable difference can be seen in Seligmann's case. In the early 1920s he searched for a reconciliation of Mendelism and behavioural characteristics. He suggested that introversion and extroversion were genetic characters following the laws of segregation and dominance and that a comparative psychology of 'primitive' and 'advanced' peoples could be built up around this notion.[20] However, this suggestion bore no fruit. But by the 1930s intelligence tests were taking the place of all the other measures of behavioural difference as the principal system around which inferiority and superiority were to be organised and it has remained so.

'Primitive' peoples had always seemed to anthropologists to 'fail' the cultural tests that they were given. Intelligence tests appeared to be one apparently measurable indication of this failure. It was easy, therefore, to assert that behind this failure lay some vague genetic determination. If this could not be identified exactly, nonetheless the reappearance of failure in succeeding generations must, it was argued, indicate inheritability and hence, eventually, genetic determination.

In addition, there was another area in which social Darwinism persisted. With the collapse of the 'comparative method', the importance of origins of human behaviour in animal life became a less important topic in social theorising. Psychology was no longer so preoccupied with tracing the germs of qualities of man in animals. However, it is also true that, once again, comparative evolution was a sub-current of psychological thought. Piaget, for example, used genetic analogies in a very similar way to Romanes and Darwin.

The study of animal behaviour had shifted to ecology, the charting of population distribution and relations between various species in an area, a kind of geo-politics of the animal world. Nonetheless, a few workers were still examining animal psychology although with a greater emphasis upon the ecological aspect and on the observation of animals in their habitat. A strong current of social Darwinism re-emerged from some work in this area, for example that of Konrad Lorenz. It could be argued that the full popularity of this revived social Darwinism arrived with another feature of intellectual life, the rejection of industrialisation and rediscovery of the importance of the natural world—a process paralleling the early twentieth century preoccupation with instinct. This stemmed from a similar feeling among a certain stratum of intellectuals of the intractability of contemporary social conflict. This trend romanticised nature and instinct, and is to some extent reflected in the popularity of books which emphasise the necessarily 'natural' character of man. But it also shared with its predecessors the discovery of the essential unchanging character of the structure of human society showing how its natural basis reflected war, acquisitiveness, property, aggression and inferiority of the female. According to the populariser of this version of social Darwinism, Robert Ardrey,

To conclude that human obsession with the acquisition of social status and material possessions is unrelated to the animal instincts for dominance and territory would be to press notions of special creation to the breaking point. To conclude that the loyalties or animosities of tribes or nations are other than the human expression of the profound territorial instinct would be to push reason over the cliff. To conclude that feminine attraction for wealth and rank and masculine preoccupations with fortune and power and fame are human aberrations arising from sexual insecurity, hidden physical defects, childhood guilts, environmental deficiences, the class struggle or the cumulative moral erosions of advancing civilisation, would in the light of our new knowledge of animal behaviour be to return man's gift of reason to its Pleistocene sources unopened.[21]

The popular writings on the 'natural' in man and society

strongly reflect the two-sided nature of this trend. Lorenz began his studies with a reinvestigation of instinct. Therefore, he began from an explicit Darwinian and nineteenth-century tradition. It led him to a kind of romantic rediscovery of the importance and even 'moral' value of the natural world and also to a highly conservative interpretation of human development. Social evolution, by taking man away from nature, had led to his degeneracy. In contrast, parts at least of the animal world appeared to demonstrate all the qualities of authenticity, monogamy, uprightness, which the human world failed to demonstrate. But this appreciation of nature produced also a strange overvaluation of certain social institutions which themselves were held to be 'natural' and legitimised by that fact.

The popularity of writing on the natural origins of man's social life was, in fact, a tribute to the combination which popular versions of animal ethology managed to convey of a romantic revolt against civilisation and, at the very same time, an almost religious reverence for certain cultural institutions. It could be asked exactly what aspects of civilisation fell under the rubric of 'unnatural' as opposed to the 'natural'. The answer was sadly unoriginal — very much taken from the eugenic catechism. They were industrialisation, the growth of cities, mass democracy and social welfare. In this respect the original early twentieth-century British exponents of instinct represented quite a different tendency from the later ones. True they regarded these institutions as necessitated by man's fall from grace. However, they also regarded man's removal from nature as an inevitable and necessary process if reason was to emerge in history. Thinkers like Wallas and Hobson believed that industrialisation was something with which societies had to come to terms and that a return to nature was a return not to authentic feeling but to barbarism.

In many ways the movement towards investigation of social systems rather than their evolutionary order represented the collapse of what the Victorians and Edwardians had been so proud of — the comparative method. The comparative method was, in their opinion, a universal means of elucidating historical development applicable in both sociology and biology. Thus the methods of biologists and those of social

evolutionists were frequently compared. Both sought historical order and to do this they examined early organic forms and
early societies respectively to find the sequence of biological
and social evolution. Their pride was to some extent justified
in so far as this did represent a cross-reference of methods.
Haddon looked for the early history of a species by examining
its embryonic development. He also looked at early human
society to find the history of his own.

But the comparative method collapsed for a number of
reasons: partly the dissatisfaction of anthropologists and social
theorists about the amount it could actually tell one. It also
died a death in biology. Garland Allen recounts in his history
of twentieth-century biology a movement away from the
preoccupation with finding historical sequence and with comparative morphology and embryology.[22] Biologists began to
doubt that these methods really told the story they were
claimed to do. Instead the physiological process of organic
forms began to be examined for their own sake rather than to
elucidate general evolutionary histories. This paralleled the
new interest in social theory in examining the actual workings
of society. The fact that social theorists were thus deprived of
an apparent biological justification for evolutionary
methodology had some effect on their attitude to traditional
social Darwinism in the twenties and thirties. The comparative method had meant that biology and sociology could share
a language and methodology. With the advent of a mathemat-
ical genetics this connection decreased although it still left an
area of social policy around which questions of biology, genetics and natural selection could be discussed.

But social Darwinism represented some very persistent and
largely unchanging ideas about society — the importance of
individual faculty in social affairs; evolution as the growth of
rationality; the belief in the superiority of contemporary social
systems; the idea that there exists a general law covering social
and natural events; the notion of the 'natural' in man and its
contribution to social structure; the importance of the natural
and animal world as a reflection of the social; the use of
analogies from nature.

What happened was that these did not disappear but the
particular form they took altered. Partly this was the result of a

fissure appearing, separating the developments in science from the particular character of the social thought which claimed a connection with it. Thus as population genetics appeared, areas popularly believed to be 'Darwinian' slipped out of sight. Social thought had to reconstitute these links. Secondly, analogy took the place of historical reconstruction as attention moved to the social system and these analogies acquired an 'empirical' content.

But lastly and significantly, the political constellation in which the social Darwinism of the late nineteenth century was created changed. In Britain with the rise of Labourism, social thought shifted more towards integrative and corporatist approaches to social problems — not completely but substantially. The emergence of Soviet Russia cut off a section of intellectuals and scientists from certain presuppositions about society common before the First World War, and the rise of Nazi Germany widened the gap. Social Darwinism became more self-conscious, more anxious to disguise its roots and more concerned with an appearance of sophistication and the acceptability of its precepts as 'theory'. To some extent these factors have succeeded since many contemporary theorists find it hard to see twentieth-century social thought as, in many cases, merely a refinement of a number of existing themes in the intellectual history of social Darwinism.

X PHILOSOPHY, SOCIAL SCIENCE AND DARWIN

ONE important characteristic of Darwinism has been its failure to satisfy the canons of scientific methodology as outlined by conventional philosophy of science. In the nineteenth century the attack on Darwin came largely from two sources. There were those referred to by Huxley as 'trained exclusively in classics and mathematics who have never determined a scientific fact in their lives and who prate loudly about Mr Darwin's method which is not inductive enough not Baconian enough forsooth for them'.[1] Darwin could not satisfy this school of philosophy because he could not offer direct observational or experimental proof of the transmutation of species. Moreover, some observations seemed to contradict natural selection. For example individuals seemed to be intermediate forms between their parents — a phenomenon dubbed 'blending inheritance'. This appeared to make it unlikely that variations would survive intact to the next generation. They would be lost or modified by the process of blending. How then could evolution based on these organic variations take place? Doubt about whether Darwinism lived up to the rules of induction regarded by many nineteenth-century philosophers as the basis of science, led even sympathetic observers like John Stuart Mill to give only a cautious welcome to the theory of natural selection. In 1860 he wrote: 'Though he (Darwin) cannot be said to have proved the truth of his doctrine he does seem to have proved it *may* be true.'[2]

However, there were other laws of knowledge which the theory of natural selection was held to have transgressed. Several of the propositions of Darwin's theory — that variation was 'random'; that species were not absolute categories but were constituted historically — were also criticised. For some philosophers knowledge was the approximation of the mind to the absolute nature or essence of things — a philosophical approach dubbed 'essentialism' by D. L. Hull

(1973). Darwinism was based on the premise that species were modifiable and had evolved from other species. To a number of philosophers — unhappy at the idea that variation and change were more than mere epiphenomena in nature — this amounted to the denial that there could be any species at all. As for the notion of 'random' variation, as late as 1929 Nordenskiöld criticised this as 'excluding the possibility of any law bound phenomenon in existence. Herein lies the greatest weakness of the Darwinian doctrine of selection'.[3]

What is the relation between Darwinism and present-day philosophy? Much contemporary philosophy has modified the strict 'inductivism' prevalent in the nineteenth century. In particular Karl Popper has popularised the notion of falsification as the criterion of scientific practice. This philosophy of science — of great influence in the English-speaking world — was developed largely in response to changes in the physical sciences.[4] Nonetheless, it also removes some of the inductivist objections to Darwin. It does not insist on immediate experimental verification of theories. In any case it regards observational 'proof' as more problematic than nineteenth-century philosophies of the inductive method. It gives hypotheses a much greater role in scientific development. Whilst it retains, in Popper's view, a strong belief in the value of verification, it does not insist on it. Popper's theory asks only that a scientific hypothesis be cast in such a way that it can *potentially* be falsified. Some students of Darwin — for example, Michael Ghiselin — have argued that, at last, Darwinism has found a philosophical defence.[5]

This is only true to a certain extent. As Popper has pointed out, natural selection could only meet the criterion of falsification if it set out a series of predictive statements which could potentially disprove the theory. However, natural selection is not primarily predictive. It largely describes the reasons for events that already have taken place.

Popper's response to this problem in the case of natural selection has been the following: 'I have come to the conclusion that Darwinism is not a testable scientific theory but a *metaphysical research programme* — a possible framework for testable scientific theories.'[6] A metaphysical research programme could be said to play the same role to the one attributed by the French historian of sciences Georges Canguilhem to vitalism in the early nineteenth century.[7] Vitalism Canguilhem argues aided the explanation and investigation of reflex action merely by detaching the investigator from absorption in the classical mechanical theory of physiology. By doing this it cleared the way for new kinds of hypotheses about reflex action. It did this without committing a scientist to an explicitly vitalist theory of reflex. In other words a non-scientific theory played an important historical role in scientific development. However, to attribute a similar function to Darwinism — for example in providing a non-theistic theory of evolution as Popper claims it has done — would be to ignore the part that natural selection plays at the heart of theories of population genetics and ecology. Natural selection has played more than a purely historical role — clearing the ground for the development of scientific hypotheses which are distinct from it. Natural selection forms an indispensable part of the present day theoretical framework of these disciplines.

These problems in philosophy of science have led to the revival of historical rather than philosophical treatments of scientific development. The most popular of these has been the theory of paradigm change associated with T. Kuhn.[8] According to Kuhn, scientific development takes place when a set of principles and experimental practices are replaced in the scientific community by another set. He suggests that what science is, basically, comes down to the ideas generally accepted by the scientific community at any one time and encapsulated in their normal practice. As his critics have pointed out, this theory may enable one to construct a social and historical account of scientific development but not necessarily an intellectual one. The history of Darwinism illustrates this. Darwinism became a paradigm of late nineteenth-century science but only in a highly restricted sense. Since a satisfactory theory of heredity was not available, the real intellectual revolution accomplished by natural selection was delayed until the 1920s. Nonetheless, Darwinism did inaugurate paradigmatic change before that time. Since most 'normal' science in biology in the latter half of the nineteenth century was laboratory inductivism, Darwinism served to give an 'historical' slant to the existing practices of comparative morphology and physiology.

Further it was fashionable to ascribe even more substantial changes to the influence of Darwin. The notion 'comparative method' was used to describe the procedure of constructing evolutionary histories in sociology, psychology, biology, history and anthropology. Many of the late Victorians believed that an intellectual revolution had affected methodologies in a wide range of disciplines. They would have been surprised to learn how little of Darwinism's true value they had, in fact, extracted and how short-lived the 'comparative method' was to be. It is not always correct to conflate apparent revolutions in 'normal' science with real intellectual changes.

Are we left, therefore, as far as Darwinism is concerned, with Paul Feyerabend's agnosticism? He argues that scientific development is 'a complex, heterogeneous historical process which contains vague and incoherent anticipations of future ideologies side by side with highly sophisticated theoretical systems and ancient petrified forms of thought'.[9] To some extent the history of Darwinism contains all these elements. It consists of a highly sophisticated theory — natural selection — together with a number of 'metaphysical' systems. Darwin's theory of the development of human faculty and society was, on the whole, patched together out of contemporary philosophical and social theory. Not much of it was logically derived from the theory of natural selection itself. But it did associate Darwinism with a powerful social metaphysic. As for the 'petrified' forms of thought referred to by Feyerabend, what characterises much of social Darwinism is the retranslation of its terms into those of the natural and moral economy of pre-Darwinian social thought. Along with the development of Darwinian science went a concomitant process in which its terms and vocabulary were incorporated into highly restrictive and ancient social aphorisms.

However, there need not be a complete philosophical anarchism about the history of Darwinism. There are some consistencies and patterns which emerge. Those which concern this book are the relationship between Darwin and philosophy, and secondly, the relationship between Darwinism and social thought. The two have been very strongly interconnected. In fact the mediation of Darwinism into social thought has been, largely, accomplished by the philosophical reinterpretation of the significance of natural selection. Michael Ruse has, recently, projected this relationship into the future:

I am absolutely convinced that, in the future, just as one end of biology is merging with the physical sciences, so at the other end biology will merge with the social sciences. More and more we shall see disciplines like psychology, sociology and anthropology incorporate into their theories results first discovered by the biologist. As this happens the philosopher will have an increasingly important role to play and I think, conversely, the meeting of the biological and social sciences may throw invaluable light on such traditional philosophical problems as those of free will and determinism and the nature and relation of mind and body.[10]

In fact philosophy has always played an important part in assimilating Darwinism into social thought and, in the process, has centred its preoccupations closely on the question of 'free will' and the 'mind–body' relationship. Even discussions of the methodology of Darwinism played a role in this. In certain cases the term 'scientific' was withheld from Darwinism not only for theoretical reasons but also because of what were considered religious, social and political implications of Darwinism. In our society to give or withhold scientific status is a powerful ideological weapon. Often when the history of Darwinism forced an admission of its role in scientific development, this was conceded only with substantial philosophical revision of its premises. Overwhelmingly the crucial area in which these revisions were demanded was the relation of the theory of natural selection to man.

II

One of the objectives of these revisions was the restoration of the importance of mind in evolution. Darwin, himself, accepted its role. In the *Descent* he suggested that mind had emerged by natural laws to become the most important factor controlling human evolution. But to many, certain ambiguities remained. Firstly, Darwin insisted that natural selection *continued* to operate at every stage of evolution. Mind had changed the course of evolution but not to the extent of rendering natural selection inoperable. Secondly, mind did not form part of the explanation of natural selection itself— it was not present in its theoretical structure.

D. L. Hull has described the philosophical revisions of the theory of natural selection as being of two sorts:

> The two modifications of evolutionary theory most frequently suggested where the substitution of evolution by saltation for Darwinian gradual evolution and the addition of a force or principle internal to organisms to direct them in their evolutionary development. The former served to salvage the discreteness of natural kinds, the latter to save teleology.[11]

This interpretation is correct but, at the same time, the purpose of these revisions was also to integrate mind in evolution. For example, to Bergson 'saltation' — change by large qualitative jumps rather than small variations — was a way of restoring a mysterious psychic force in evolution. Some took up the idea that Darwinism had shattered the notion that species were absolute to argue that species had no real existence but were merely categories of thought. This made 'mind' the principle of order and development in the world. Reality was merely unreconstructed chaos. The notion of randomness or contingency was applied to a similar end. William James argued for a theory of psychological evolution in which 'free will' and 'choice' provided the 'random' elements on which natural selection worked. Before William James, Stephen and Bagehot gave this idea a sociological perspective. Society progressed by the free exercise of rationality embodied concretely in invention and innovation.[12]

Charles Peirce who understood better than any other philosopher of the nineteenth century that Darwinism was essentially a science of probabilities used the indeterminism this implied in even more complex ways. He suggested a variation on Paley's watch. It was not the ordered structure of the watch that implied an intelligent maker but the combinations of chance occurrences which, in the long run, produced the marvellous features of an adapted and ordered universe. He saw in randomness and chance, because it ultimately produced adaption and structure, a higher rationality than that suggested by the simple mechanical principles in Paley's watch.[13] Bergson also used the idea of contingency to restore the laws of mind. He suggested, for example,

Suppose also that the idea of disorder arises in our mind whenever seeking one of the two kinds of order, we find the other. We may say that this first kind of order is that of the *vital* or *willed*, in opposition to the second, which is that of the *inert* and *automatic*.[14]

Most of these theories were either reflections on, or in Bergson's case, rejections of Darwinism. In addition to these, there were attempts at revising the theory of natural selection itself. G. J. Romanes and J. M. Baldwin both undertook revisions in which mind and free will would be given greater importance. One of the latest is that of Karl Popper. Popper's aim is to show how 'non physical things (such) as *purposes, deliberations, plans, decisions, theories,* and *interpretations* and *values* can play a part in bringing about physical changes in the physical world.'[15]

His argument is that choice and reason can provide the framework in which genetic change occurs. Thus they select the appropriate physical variation by determining the environment or behaviour pattern in which variation will occur. Those most appropriate to the new environment or behaviour will be selected.

Once a new aim or tendency or disposition or a new skill, or a new way of behaving has evolved in the central propensity structure, this fact will influence the effects of natural selection in such a way that previously unfavourable (though potentially favourable) mutations become actually favourable if they support the newly established tendency.[15]

This, in Popper's view, leads to a simulation of the Lamarckian or Bergsonian notion of a psychologically directed evolution without actually abandoning natural selection.

When A. R. Wallace rejected a similar compromise suggested to him by Baldwin, he did so for several reasons. The major one was its irrelevance. Unless a direct link between a 'preference' and a genetic change which suits the preference could be discovered, then variation occurs outside of and, often, in opposition to our preferences. The most that can be said is that one of the factors in natural selection is the environment or ecology in which an individual exists. We may choose to alter this or inhabit another. But the randomness or contingency of variation would not be solved thereby.

The desire to make Darwinism conform to 'free will' is of

little value to the theory but it has a great many implications for social thought. Mivart believed that a psychological theory of evolution would preserve the notion of God's intervention in the physical world. In a different context Popper believes that such a modification of natural selection would preserve man's capacity for intervention in evolution. It would avoid a determinism in which, 'the whole world with everything in it is a huge automaton and that we are nothing but little cog wheels or at best subautomata within it . . . .'[15]

Views such as these often seem to be linked to the notion of human freedom although socialist versions of them (for example Lysenkoist biology — which to use the socialist term was 'voluntarist' in its view of social progress and which attributed a very vigorous psychic life to plants and animals) are paradoxically often seen as philosophies of human manipulation. This capacity to see the exercise of 'free will' as desirable in one political context and manipulative in another can be explained by another feature of these doctrines. Most of the theories of 'free will' — including Lysenkoism — rely strongly upon a particular kind of delineation of human essence. Thus most theories which criticise Darwinism for the failure to include choice, will and preference have, in fact, a very tightly drawn picture of the 'nature' which exercises these choices and the range of options open to it. Mivart, for example, quite clearly believed in an unchanging human essence. As Bergson put it he looked at disorder to discover another kind of order — in his case a divinely inspired one. Thus although Darwinism was 'deterministic', according to Mivart, it led to social chaos and to 'horrors worse than the Parisian Commune'.[16] Thus Mivart's theory of free choice led to quite a rigid social order. It did not allow any of those kind of choices out of which Parisian Communes emerge.

Similarly once mind is released in evolution, history can become highly determined. Popper, for example, sees an historical order which, by the criterion of intellectual creativity, creates a hierarchy. Natural selection, as Darwin described it, operates among the lower reaches of life. In our own society we escape these natural laws, but,

It is different with primitive man and the amoeba. Here there is no critical

attitude and so it happens more often than not that natural selection elimi-
nates a mistaken hypothesis or expectation by eliminating those organisms
which hold it or believe in it.[17]

To a great extent Popper's liberation of mind from evolutionary laws and the subjection of evolution to mind brings us
full circle to social Darwinism. To treat philosophy's interest
in Darwin at its own valuation as though it was purely a
methodological reflection on the theory is to ignore the strong
connections it has with social theory. One of the earliest social
Darwinisms — that of Bagehot and Stephen — was an application to society of the notion of free choice and especially of
rationality in human social evolution. Bagehot even toyed
with a version of Darwinism which gave behaviour a major
role in determining variations. They too found a hierarchy of
rationality in history and, in spite of their insistence on the
freedom of the intellect, were highly conservative in their
attitude to certain social institutions. This dialectic between
'free will' and conservatism is thrown into sharp contrast in
modern social Darwinism — especially in certain aspects of
contemporary sociobiology. Many of the extrapolations to
human social development from sociobiology show the per-
vasiveness of those philosophical themes.

III

In the 1970s a new series of 'Darwinian' reflections on human
society have emerged largely due to the impact of advances in
animal ethology in post-war years. Several animal ethologists
claim that the principles of what has become known as
sociobiology can be applied to human social institutions.
These investigations of human social behaviour cover a wide
range. Some make modest claims about what sociobiology
can reveal about human society. Others — in particular the
work of E. O. Wilson — make much greater ones. Wilson
considers that sociobiology is on the verge of revolutionising
the more traditional disciplines which investigate human society and behaviour. These include not only sociology but also
psychology, ethics and even history.

It may not be too much to say that sociology and the other social sciences as well as the humanities are the last branches of biology waiting to be included in the Modern Synthesis. One of the functions of sociobiology, then, is to reformulate the foundations of the social sciences in a way that draws these subjects into the Modern Synthesis. Whether the social sciences can be truly biologised remains to be seen.[18]

These thinkers consciously distance themselves from what they regard as the social Darwinism of the past. They tend, wrongly, to see it as purely a theory of competitive individualism and racial conflict. By defining it in this way, they are, in fact, disguising from themselves their own heritage. Even the most pacific of contemporary sociobiologies shares many concepts with the social Darwinism of the late nineteenth and early twentieth century. In discussing human society, sociobiologists have taken up many of the notions characteristic of traditional social Darwinism. They have also returned to a number of the philosophical ideas outlined above. This has the curious effect, as we shall see, of bringing together E. O. Wilson, a natural historian, and Karl Popper, a philosopher, at the point at which they discuss the relationship between biology and human social institutions.

What is new in the animal ethology of the post-war period is its use of population genetics. In particular animal social structures have been investigated to discover the relationship between them and the 'inclusive fitness' of their members — that is the relationship between social structure and differential fertility. By working out the cost benefit in genetic terms for the individual of the various forms of animal social structures, explanations can be found for a wide range of behaviour which might, at first, appear paradoxical from the point of view of natural selection. The most famous demonstration is the genetic cost benefit of altruism. Sociobiologists have given a new meaning to altruism in the sense that they have revealed it not as a sentiment for the survival of the group or species but a strategy which may lead to a greater survival chance for the individual's own genes. This, they argue, can explain the survival value of seemingly self-sacrificing behaviour. These theories are truly in the Darwinian tradition in their emphasis upon differential fertility and upon the survival of certain hereditary variations within a population.

Two points remain to be made which sociobiologists bring out clearly in the case of animal society but often are less clear about in their explanations of human social life. One is that animal society is an optimum population within a particular ecological niche. The conditions for differential fertility change because of weather, changes in resources etc. In addition the onset of seasons trigger off behaviour change in these societies. In other words animal societies are always interactions between those individuals who comprise it and external factors. The second is the great variety of animal social structures to which this gives rise. There are variations even between closely allied species and sometimes within species.

Contemporary sociobiologists are not the only ones who have tried to see the relevance of these ideas to human society. Nineteenth-century theorists also examined the relationship between culture and differential fertility. Although they had no knowledge of population genetics they did see that Darwinian 'fitness' was measured by the number of descendants an individual left and they speculated about the effects of the social structure on this. Galton, for example, examined the relationship between differential reproduction and the competition for economic resources in his own society and he observed a curious feature of it. Whereas the conventional social Darwinism of his day hypothesised that control of resources was the result of superior 'fitness' there seemed to be a negative correlation between wealth and numbers of descendants. For example, Galton observed the paradox of the heiress. Socially sought after, economically desirable, she often came from a family with a relatively low fertility. Though she was almost certain to marry, to marry an heiress was often to risk diminished fertility. Worse than this, in Galton's opinion, was the price paid by the professional middle class for social status and occupational success. This price was often delayed marriage. It consequently represented an exchange of fertility for social status. Finally there seemed to be a negative correlation between resources and reproduction. The poor were numerous and though later eugenists came to the conclusion that complete destitution often reduced fertility, nonetheless those with relatively little seemed to leave more descendants than those who had a great deal.

It was as if wealth itself had taken over the role of the 'selfish gene'. It maximised itself by a series of social strategies — laws of inheritance, prejudice against marriages which interfered with the process of its accumulation — many of which rewarded reduced fertility. Wealth reproduced itself rather than people and especially it *differentially* reproduced itself.

When Wallace arrived in the Malay Archipelago he encountered similar problems. He was interested in the question of human ecology — why and where human populations dispersed and the relationship between fertility and social structure. If Malthus was correct and human population expanded to meet the available resources then, Wallace felt, certain areas in the Archipelago ought, because of their natural resources, to support a much larger population than they in fact did. Many of the peoples he met had a technology often beautifully adapted to the exploitation of resources given their particular social organisation. But they were also often poor with meagre diets and high mortality. They lacked the social means by which potential abundance could be transformed into actual abundance. These speculations by Wallace show how far, in spite of his utilisation of the Malthusian notion of the effect of expanding population on selection, he had a very real appreciation of the limits of Malthusianism as an explanation of the basis of that selection.

Finally in the twentieth century the population geneticist J. B. S. Haldane also considered the relation between differential fertility and social structure. Modern animal ethologists remember his speculations upon the genetic value which would accrue to an individual in a seemingly altruistic act — for example saving a relative from drowning. Calculations of the genetic advantage of such 'altruism' form a major basis of sociobiology. However Haldane also speculated on the effect geographical and social isolation would have on human populations. The latter factor — social isolation — was not always the product of the former but was also imposed on human populations by the exigencies of religious, racial and class divisions. In other words human societies might — unconsciously as well as consciously — *create* kinship groups or exclusive strata which will affect their genetic make-up and their differential fertility.

What distinguishes these thinkers is not only that they had thought about social structure and differential fertility but also that they had encountered the problems which arose by thinking of social institutions purely as a function of it. This might be because — as in Galton's case — they disapproved of the effects social institutions were having on the relative fertility of social classes. (This opinion has resurfaced in some sociobiological literature.) But in the other cases, Wallace and Haldane recognised and accepted that culture would be governed by laws not reducible to biology and yet that these laws would also play a part and intervene in the patterns by which the human species reproduced itself. To elucidate the point at which these two factors intersect is potentially no more complex than identifying the relationship between animal habitat and reproduction. But the laws of human social evolution are a great deal more complex as both Wallace and Haldane saw.

Unfortunately much contemporary sociobiology has retreated from this in examining human society. It has done so largely by introducing into the work on human evolution a basically philosophical proposition. This is the idea that human society can be explained by reducing it to a function of individual behaviour. This, as Comte and Mill saw, was the main basis upon which social theory laid claim to biology. The result has been the re-emergence of very much the same difficulties as perplexed a previous generation of social theorists. It has also, in the case of E. O. Wilson, produced a revision of biology very similar to that carried out by philosophers of 'mind' and free will.

The opinion of Jerome Barkow that,

Human social institutions are and can only be patterned expressions of biologically based learning preferences, predispositions and motivations, characteristics which are there because they once at least maximised inclusive fitness.[19]

is very similar to the proposition put forward by Auguste Comte in the first half of the nineteenth century that,

Biology will be seen to afford the starting point of all social speculation in accordance with the analysis of the social faculties of Man and the organic conditions which determine its character. But . . . we must construct them by applying the positive theory of human nature to the aggregate of corresponding circumstances.[20]

This is an old theme in social thought and its consequences for analysis of social behaviour can be seen in E. O. Wilson's treatment of the economy. The sociologist Max Weber believed that man saved and capitalism appeared. The economist Alfred Marshall believed that capitalism arose out of thrift and economic class distinctions were the product of the distribution of thrift, the endurance of steady labour and other moral psychological faculties.Similarly Wilson quotes from those economists who believe that economic activity rests on 'motivation, *esprit*, effort, persistence, and other psychological variables'.[21] He also sees in 'primitive' society the development of the psychological distinctions which Alfred Marshall thought were so important in generating economic classes:

> With a small tribe of 'Kung Bushmen' can be found individuals who are acknowledged as the 'best people' . . . . There are some who never try to make it and live in run down huts and show little pride in themselves or their work.[21]

The connection of theories like these with psychological hierarchies of race and class has been examined in a previous chapter. In one way, theories of this kind simplify social theory — over-simplify would be a better word — since they invariably spirit away the complexity of social relationships. But they also cause problems for a biological sociology for they frequently require a theory of organic evolution based on the same premise of individual adaptive effort. As Wilson's critics have pointed out, human culture changes at a faster rate than genes do. This would lead one to suspect that genes and social behaviour are not so closely interconnected as many sociobiologists claim. Wilson avoids this problem in two ways. First of all he uses the term phenotype to describe cultural behaviour which is not genetically determined. Phenotypes are non-inheritable characteristics which may, nonetheless, enhance survival. Secondly, he suggests that the genetic basis for behaviour may not be individual genes for particular behavioural patterns but a genetic basis for a general ability to create culture and learn social rules. However, as Stephen saw, a non-inheritable social characteristic has to rely on social not biological mechanisms for its survival. But if Wilson granted culture its own strategies for reproducing itself, he would have severely limited the importance of biology. Therefore he argues that cultural phenotypes are 'tracked' by genetic evolution. Eventually the changes in a social system will produce a personality which fits its institutions and has a genetic base.

This solution recalls Bagehot's theory of national character produced by the mechanism of use-inheritance. It is also very close to Popper's revival of the 'Baldwin effect'. In fact it is a departure from the principles of neo-Darwinism upon which the most original advances of sociobiology ultimately rest. The justification for this advanced by Popper was to restore mind and free will in evolution. As we shall see, Wilson's theory of society and culture rests upon very much the same premises.

One might argue that it is precisely because culture can change so fast or be transformed drastically within a comparatively brief time, that evolution would favour a general adaptability rather than an hereditary basis for any particular social institution. This would, however, remove the investigation of human social behaviour largely out of the hands of sociobiology. Wilson is searching less for a scientific theory of human social evolution than for an ontology of human nature in which there is a strong biological component.

As Wilson's recent work shows, he is more concerned with ethics than with social theory. The insights of animal ethology are, in his own words to be 'a pre-requisite for the creation of a fully rationalistic morality'.[22] Thus Wilson is basically searching animal ethology for ideas about the limits of human moral freedom. In addition he is trying to construct a typology of human nature around which these moral choices are structured. Like all such theories it is achieved by the suppression rather than the elucidation of social processes. It resolves social theory into a philosophy of human individuality and choice. When culture emerges it does so only as an effect of individual activity or as an obstacle to it.

Many of his explanations of the relationship between man

and culture recall those of the social theorists of instinct in the early twentieth century. In particular the notions of culture as a drag upon man's aspiration towards perfect freedom and, also, the notion of human nature as a form of historical punishment. Since society is always reducible in Wilson's view to questions of individual behaviour, then his theory of human society has almost totally been submerged into one of individual moral choice. It becomes, therefore, impossible for him to ask any of the questions about social structure and differential fertility that were asked by Wallace, Haldane or even Galton. A scientific theory of society was promised by him but he has offered instead a morality. In many ways, the philosophy traditionally surrounding Darwinism offered precisely the same thing.

IV

This translation of biological evolution and particularly Darwinism into moral theory has been a constant feature since the *Origin*, first appeared. Bagehot replied to Henry Buckle's belief in the importance in evolution of the material arts and climate by arguing against

Mr Buckle's idea that material forces have been the mainsprings of progress and moral causes secondary and in comparison not to be thought of; on the contrary; moral causes are the first here. It is the action of the will that causes the unconscious habit; it is the continual effort of the beginning that creates the hoarded energy of the end; it is the silent toil of the first generation that becomes the transmitted aptitude of the next. Here physical causes do not create the moral but moral create the physical; here the beginning is by the higher energy, the conservation and propagation by the lower.[23]

This theme is a constantly repeated one in social Darwinism without necessarily the explicit commitment to use-inheritance evident in Bagehot's work. Social Darwinism is built around the idea that the exercise of individual faculty explains society. It also rests on the idea that the social order corresponds to the moral. Finally there is a strong element in social Darwinism which insists that biological laws themselves should conform to these expectations, that nature should show the operation of individual consciousness and, in some theories, should actually reflect the moral order.

Because of this, those who see Darwinism as ultimately a highly conservative doctrine may have, in fact, identified a real characteristic. However many variations of social theory it has produced, social Darwinism implies that individuals are allotted social places through their heredity or their moral choice. In the first case this means that the social places we occupy are inevitable; in the second, that we deserve them. This relationship between faculty and society can be discussed in a highly abstract way but the practical implications of it in social life do not lend themselves to a flexible view of, for example, the reasons for social hierarchy. These conservative implications emerged very quickly. So did its continuity with certain religious ideologies of social order. The social functions performed by these ideologies were, to some extent, taken over by 'scientific' theories of how social roles are distributed.

It is possible to produce an 'oppositional' or 'radical' view of the role of human nature in society, to say, for example, that all human beings are by nature equal and capable of exercising moral choice. Mary Midgley has argued that theories of freedom require the biological limits of human nature to be built into them; that this by no means results in a diminution of human social freedom but only a surer foundation on which to build it.[24]

A theory of this sort in a social context, however, can only justify notions of social change by arguing that they make society conform better to man's nature. Questions of what this nature is are very seldom — in spite of appearances to the contrary — settled at the level of abstract argument. Social institutions tend to dictate the limits and character out of which an 'abstract' human nature is constituted. The human 'nature' Darwin was required to explain in the *Descent* was a composite of the social and political conceptions, about how Victorian men and women of his time and social position behaved. He harnessed various features of contemporary intellectual life to explain it. In the process he helped begin the creation of social Darwinism. For ultimately social Darwinism resolved itself into a theory which began from the 'reality' of existing social relationships — or more often an ideological picture of them — and argued back to their 'natural' causation. This led to a methodology not simply of reductionism to biology but one which could not escape the limits of the social ideologies of its time.

Therefore, in spite of the real revolution in the biological science carried out by Darwin and Wallace, social Darwinism was unable to effect a similar revolution in social theory. It represented a continuity with previous social thought in certain important respects. The search for a social theory was, for the vast majority of nineteenth-century sociologists, a search for a 'natural' underpinning to social order and in addition for a theory of the individual's obligation to respect that order. Comte who in many ways inaugurated an era of methodological self-consciousness on the part of social theorists should therefore be given the last word on the general project of sociology. He praised 'De Maistre's fine political aphorism "Whatever is necessary exists" ', thus committing social theory *and* nature to a teleology from which it has seldom escaped. His comment illustrates how likely it was that social theorists would search for a reflection in the natural world of the kind of order they imposed upon the social. Like Bagehot, Comte insisted that this order was moral and intellectual '. . . the intellectual and moral condition of human existence are as real and as imperative as its material conditions'. Finally he set out the limits of social theory: 'No sociological view', Comte argued, 'can therefore be admitted . . . that is contradictory to the known laws of human nature'.[25] The object of this book has been to show how these 'known laws' were remarkably fluid and how their constitution was an historical and social process rather than a scientific one. It has also been to show how the search for them produced tension and contradiction between social theory and biology.

SELECT BIBLIOGRAPHY

Before 1920

Place of publication is London, except where otherwise stated. Included in this section are works of an author after 1920 if the majority of his books relate to the preceding period.

Alexander, Samuel, *Moral Order and Progress*, Trubner, 1889.

Allen, Grant, *Charles Darwin*, Longmans, Green, 1885.

Angell, J., *The Influence of Darwin on Psychology, Psychological Review*, Vol. XVI, 1909.

Anon, 'Darwinism and National Life', *Nature*, Vol. I, Dec. 1869.

Aveling, Edward, *Darwinism and Small Families*, Besant and Bradlaugh 1882.

    'The Gospel of Evolution', *The Atheistic Platform*, No. 3, 1884.

    The Religious Views of Charles Darwin, Freethought Publishing Co., 1883.

    The Student's Darwin, Freethought Publishing Co., 1881.

    'Charles Darwin and Karl Marx', *New Century Review*, Vol. I, 1897.

Bagehot, Walter, The English Constitution, Chapman & Hall, 1867.

    Physics and Politics, King, 1872.

Bain, Alexander, *Mental and Moral Science*, Longmans, 1868–72.

    On the Study of Character including an Estimate of Phrenology, London, Parker Son & Bourn, 1861.

    The Emotions and the Will, Parker, 1859.

    The Senses and the Intellect, Parker, 1855.

Baldwin, J. M. *Between Two Wars 1861–1921*, 2 vols. The Stratford Company, Boston 1926.

    Darwin and the Humanities, Review Publishing Co., Baltimore, 1909.

    Development and Evolution, MacMillan, New York, 1902.

    The History of Psychology, Watts, 1913.

    Mental Development in the Child and the Race, MacMillan, New York, 1895.

Barker, Ernest, *Political Thought in England, 1848–1914*, Williams & Norgate, 1915.

Barrington Russell, E., *The Works and Life of Walter Bagehot*, Longmans, (10 vols), Green, 1915.

Bateson, William, *Biological Fact and the Structure of Society* (Herbert Spencer Lecture), Clarendon Press, Oxford, 1912.

    'Common Sense in Racial Problems', Galton Lecture, 1919.

    'Evolution, Faith and Modern Doubts' (An Address to the American Association for the Advancement of Science), in Beatrice Bateson, *William Bateson, F.R.S., Naturalist. His Essays and Address together with a Short Account of his Life'*, Cambridge University Press, Cambridge, 1928.

'Heredity and Variation in Modern Thought', in *Darwin and Modern Science*, (ed. A.C. Seward), Cambridge University Press, Cambridge, 1909.

Mendel's Principles of Heredity, Cambridge University Press, Cambridge, 1902.

Review of the Evolution Theory of A. Weismann, *The Speaker*, Vol. XII, 24 June 1905.

Bax, Ernest Belfort, *Problems of Man, Mind and Morals*, Grant Richards, 1912.

The Religion of Socialism, Swann & Sonnenscheinn, 1887.

Beddoe, John, *The Races of Great Britain, A Contribution to the Anthropology of Western Europe*, J. W. Arrowsmith; Bristol, Trubner, 1885.

Bergson, Henri, *Creative Evolution* (trans. 1911), MacMillan, 1960.

Bosanquet, Bernard, 'Socialism and Natural Selection', in *Aspects of the Social Problem*, MacMillan, 1895.

Brett, G. S., *A History of Psychology* (1912), G. H. Allen & Co., 1912.

Buckle, H. T., *The History of Civilisation in England*, (2 vols), Parker, 1857.

Butler, Samuel, *Evolution, Old and New*, Hardwicke & Bogue, 1879.

Life and Habit, (1877), Trubner, 1878.

Luck or Cunning, (1886), Trubner, 1887.

Cairnes, J. E. 'The Negro Suffrage', *MacMillan's Magazine*, Vol. XII, 1865.

Carpenter, W. B., 'On the Hereditary Transmission of Acquired Psychical Characteristics', *Contemporary Review*, Vol. XXI, 1873.

Nature and Man, Kegan Paul, 1888.

Carr, Herbert, W., *Henri Bergson — The Philosophy of Change*, T. C. & P. E. Jack, 1912.

Castle, W. E., *Genetics and Eugenics*, Harvard University Press, Cambridge, Mass., 1916.

Cecil, H. M., *Pseudo-Philosophy at the end of the Nineteenth Century — an Irrationalist Trio*, University of London Press, 1897.

Chambers, Robert, *Vestiges of the Natural History of Creation*, 1844; Leicester University Press, 1969.

Clifford, W. K., *Lectures and Essays* ed. Leslie Stephen and Frederick Pollock, MacMillan, 1879.

Clodd, Edward, *Pioneers of Evolution from Thales to Huxley*, Grant Richards, 1897.

Coker, Francis William, *Organismic Theories of the State* (Studies in History, Economics and Public Law), Columbia University Press, New York, 1910.

Comte, A., *The Positive Philosophy of Auguste Comte*, (2 vols), trans. and cond. by Harriet Martineau, Chapman, 1853.

Crawfurd, John, 'The Plurality of the Races of Man', a discourse delivered by John Crawfurd at St Martin's Hall, 13 Jan. 1867.

'On the Theory of the Origin of Species by Natural Selection' *Transactions of the Ethnological Society of London*, Vol. VII, 1869.

Creighton, J. E., 'Darwinism and Logic', *Psychological Review*, Vol. XVI, 1909.

Crozier, J. B., *History of Intellectual Development on the Lines of Mental Evolution*, Longmans, Green, 1897.

Sociology Applied to Practical Politics, Longmans, Green, 1911.

Cunningham, J. and Haddon, A.C., 'The Anthropological Institute of Ireland' in *Journal of the Anthropological Institute*, Vol. XXI, 1891–2.

Darwin, Charles, Appendix on 'Instinct' in G. J. Romanes, *Mental Evolution in Animals*, 1883. Kegan Paul & Co.,

'Biographical Sketch of an Infant', *Mind*, Vol. II, 1877.

The Expression of the Emotions in Man and Animals, John Murray, 1872.

The Life and Letters of Charles Darwin, ed. F. Darwin, John Murray, 1887.

More Letters of Charles Darwin, (2 vols), ed. F. Darwin, John Murray, 1903.

'The Origin of Certain Instincts', *Nature* Vol. VII, 1873.

On Origin of Species by Means of Natural Selection, John Murray, 1859. Penguin Edition ed. by J. W. Burrow, 1968. Reprinted 1970.

The Autobiography of Charles Darwin, ed. F. Darwin, Watts, 1929.

The Descent of Man and Selection in Relation to Sex, (2 vols), John Murray, 1871.

Collected Papers (2 vols), ed. Paul H. Barrett, University of Chicago Press, Chicago, 1977.

Davies, M. M. *Psychological Interpretations of Society*, Columbia University Press, New York, 1909.

Dendy, A., *Outlines of Evolutionary Biology*, Constable, 1912.

The Biological Foundations of Society, Constable, 1924.

Dewey, John, *The Influence of Darwin on Philosophy and Other Essays in Contemporary Thought*, Henry Holt, New York, 1910.

Dodson, G. R., *Bergson and the Modern Spirit*, Lindsey Press, 1914.

Driesch, Hans, *The Science and Philosophy of the Organism* (Gifford Lectures 1907), (2 vols.), University of Aberdeen, 1908.

The History and Theory of Vitalism (1905 Leipzig), trans. and rev. 1914 by C.K. Ogden, MacMillan, 1914.

Drummond, Henry, *The Lowell Lectures on the Ascent of Man*, Hodder & Stoughton, 1894.

Duncan, David (ed.), *The Life and Letters of Herbert Spencer*, Methuen, 1908.

Dunlap, Knight, 'Are there any Instincts?', *Journal of Abnormal Psychology*, Vol. XIV, 1919.

Elliot, Hugh S., 'Modern Vitalism', *Bedrock*, Vol. I, 1912.

Modern Science and the Illusions of Professor Bergson, Longmans, 1912.

Ellis, Henry Havelock, 'Genius and Stature', *Nineteenth Century*, Vol. XLII, 1897.

'The Problem of Race Regeneration' in *New Tracts for the Times* (series), 1912.

Review of Bebel's *Women and Socialism*, in *Today*, Vol. II, 1884.

Selected Essays, Dent, 1936.

Views and Reviews — a Selection of Uncollected Articles 1884–1932 Desmond & Harmsworth, 1932.

Ellwood, C. A., 'The Influence of Darwin on Sociology', *Psychological Review*, Vol. XVI, 1909.

Engels, F., *Anti Duhring* 1878, Lawrence & Wishart, 1969.

The Dialectics of Nature (written 1873–86), Progress Publishers, Moscow, 1972.

'The Part Played by Labour in the Transition from Ape to Man' (written 1876). First published in *Neue Zeit* (1895–6) in Marx/Engels, *Selected Works*, Lawrence & Wishart, 1970.

Farrar, F. W., 'Philology and Darwinism', *Nature*, Vol. I, 24 March, 1870.

Ferri, Enrico, *Socialism and Positive Science — Darwin, Spencer Marx* trans. from the French edition of 1896 by Edith Harvey, The Socialist Library, 1905.

Freeman, E.A., *Comparative Politics*, 1874. (The Rede Lecture given before the University of Cambridge, 29 May 1872).

Galton, Francis, *English Men of Science — Their Nature and Nurture*, MacMillan, 1874.

Hereditary Genius, MacMillan 1869.

Inquiries into Human Faculty, Macmillan, 1883.

Narrative of an Explorer in *Tropical South Africa*, 1853.

Natural Inheritance, Macmillan, 1889.

Memories of My Life, Methuen, 1908.

'The Possible Improvement of the Human Breed Under the Existing Conditions of Law and Sentiment', Huxley Lecture at the Anthropological Institute, 1901.

Probability — the Foundation of Eugenics (Herbert Spencer Lecture), Clarendon Press, Oxford, 1907.

'On Instruments for (1) Testing Perception of Differences of Tint and for (2) Detecting Reaction Time', *Journal of the Anthropological Institute of Great Britain and Ireland*', Vol. XIX, 1889–90.

'Retrospect of Work done at my Anthropometric Laboratory at South Kensington', *ibid.*, Vol. XXI, 1891–2.

'A New Instrument for Measuring the Rate of Movement of Various Limbs', *ibid*, Vol. XX, 1890–1.

'The Patterns in Thumb and Finger Marks,' *ibid*, Vol. XX, 1890–1.

'Hereditary Talent and Character', Part I and Part II, *Macmillan's Magazine*, Vol. XXII, 1865.

'A Theory of Heredity', *Contemporary Review*, Vol. XXVII, 1875. (Reprinted from *Journal of Anthropological Institute*).

'The Part Played by Religion in Human Evolution, *National Review*, Vol. XXIII, 1894.

'Restrictions in Marraige', *Sociological Papers*, 1905.

'Eugenics — it's Definition, Scope and Aims', *Sociological Papers*, 1905.

'Studies in National Eugenics', *Sociological Papers*, 1906.

'Hereditary Improvement', *Fraser's Magazine*, Vol. III, New Series, 1873.

Geddes, Patrick, 'On the Conditions of Progress of the Capitalist and of the Labourer' in *The Claims of Labour*, Co-operative Printing Company, Edinburgh, 1886.

'On the Application of Biology to Economics', *British Association Meeting*, Vol. LV, Section F. 1885.

John Ruskin — Economist, William Brown, Edinburgh, 1884.

An Analysis of the Principles of Economics (read before the Royal Society of Edinburgh 1884), Williams & Norgate, 1885.

Gissing, George, *The Odd Women* (1893), W. H. Norton, New York, 1971.

Grant, Sir Alexander, 'Philosophy and Mr Darwin', *Contemporary Review*, Vol. XVII, May 1871.

Greg, W. R., *Essays on Political and Social Science*, 1853.
'On the Failure of Natural Selection in the Case of Man', *Fraser's Magazine*, Vol. LXXVIII. 1868.

Haddon, A. C., *The Study of Man*, John Murray, 1898.
A History of Anthropology, Watts, 1934.
'Reports of the Cambridge Expedition to the Torres Straits' (ed.) in *Physiology and Psychology*, Vol. II, Cambridge University Press, Cambridge, 1901.
The Practical Value of Ethnology (Conway Memorial Lecture), Watts, 1921.

Hadley, A. T., 'The Influence of Charles Darwin upon Historical and Political Thought', *Psychological Review*, Vol. XVI, 1909.

Haldane, John, 'Vitalism', *Nineteenth Century*, Vol. XLIV9 1898.

Hardie, J. Keir, *From Serfdom to Socialism*, George Allen, 1907.

Harrison, Frederic, *Memories and Thoughts*, Macmillan, 1906.

Harvey, Charles, *The Biology of British Politics*, Swan Sonnenschein, 1904.

Haycraft, J. Berry, *Darwinism and Race Progress*, Swan Sonnenschein, 1895.

Hill, G. D. *Heredity and Selection in Sociology*, A. & C. Black, 1907.

Hobhouse, L. T., *Liberalism*, Williams & Norgate, 1911.
Democracy and Reaction, Fisher Unwin, 1904.
Mind in Evolution, Macmillan, 1901.
Social Development — its Nature and Conditions, Allen & Unwin, 1924.
Social Evolution and Political Theory, Columbia University Press, New York, 1911.

Hobhouse, L. T., *The Labour Movement*, Fisher Unwin, 1893.
The Rational Good, Henry Holt, New York, 1921.
Development and Purpose (1913), 2nd edn., Macmillan, 1927.
'The Diversions of a Psychologist', *The Pilot*, Vol. V. no. 97, 1902.

Hobson, J. A., *Imperialism: A Study*, James Nisbet & Co., 1902.
The Psychology of Jingoism, Grant Richards, 1901.

Hunt, J. 'On the Application of the Principle of Natural Selection in Anthropology', *Anthropological Review*, Vol. IV, 1866.

Hunter, W. S., 'The Modification of Instincts from the Standpoint of Social Psychology', *Psychological Review*, Vol. XXVII, 1920.

Huxley, Leonard, (ed.), *Life and Letters of Thomas Henry Huxley*, (3 vols), Macmillan, 1903.

Huxley, T. H., *Evidence as to Man's Place in Nature*, William & Norgate, 1872.
Administrative Nihilism, Appleton, New York, 1872.
Evolution and Ethics (Romanes Lecture), Macmillan, 1893.
'The Struggle for Existence — a Programme', *Nineteenth Century*, Vol. XXIII, Feb. 1888.
Social Diseases and Worse Remedies, Macmillan, 1891.

'Mr Darwin's Critics, *Contemporary Review*, vol. XVIII, 1871.

Lay Sermons, Addresses and Reviews, Macmillan, 1870.

James, William, *Principles of Psychology*, (2 vols) Macmillan, 1890.

Jevons, F. B., *Evolution*, The Churchman's Library, 1900.

Jevons, W. Stanley, 'A Deduction from Darwin's Theory', *Nature*, Vol. II, 1869.

Kantor, J. R., 'A Functional View of Human Instinct', *Psychological Review*, Vol. XXVII, 1920.

Kidd, Benjamin, *Principles of Western Civilisation*, Macmillan, 1902.

Social Evolution, Macmillan, 1894.

The Control of the Tropics, Macmillan, New York, 1898.

The Science of Power, Methuen, 1918.

Individualism and After, (Herbert Spencer Lecture), Clarendon Press, Oxford, 1908.

'Sociology', *Encyclopedia Britannica*, 1911.

Review of *Evolution and Religion* by E. J. Dadson in *International Journal of Ethics*, Vol. IV, 1893.

Interview in *Daily Chronicle* 20 June, 1894.

'A Note on Mr Galton's View', *National Review*, Vol. XXII, 1894.

Kirby, Reverend W., 'On the Power, Wisdom and Goodness of God, as manifested in the Creation of Animals in their History, Habits and Instincts', *The Bridgewater Treatises*, Pickering, 1835.

Knox, Robert, *The Races of Men — a Fragment* (2 vols), 1850–62.

Kropotkin, P., *Mutual Aid*, Heinemann, 1902.

'The Direct Action of Environment on Plants', *Nineteenth Century*, Vol. LXVIII, 1910.

Lafargue, P., Mr Herbert Spencer's *The Coming Slavery*', *Today*, Vol. I, 1884.

Lankester, E. Ray, *Science from an Easy Chair*, Methuen, 1910.

Laski, H. J. 'The Scope of Eugenics', *Westminster Review*, Vol. LXXIV, 1910.

Lebon, G., *The Psychology of Peoples* (Paris, 1894), Fisher Unwin, 1899.

The Psychology of Socialism, Fisher Unwin, 1899.

The Crowd (Paris 1895), Fisher Unwin, 1896.

Letourneau, C., *Property: Its Origin and Development*, Walter Scott, 1892.

Lewes, G. H., 'The Spinal Cord as the Centre of Reflex', *Nature*, Vol. VIII, 1873.

Lubbock, J., *The Origin of Civilisation and the Primitive Condition of Man*, Longmans Green, 1870.

Prehistoric Times, Williams & Norgate, 1865.

'On the Origin and Early Condition of Man', *British Association Report*, Vol. XXXVII, 1867.

MacKintosh, R., *from Comte to Benjamin Kidd*, Macmillan, 1899.

Maitland, F. W., *The Life and Letters of Leslie Stephen*, Duckworth, 1906.

Mallock, W. H., *Aristocracy and Evolution*, A. & C. Black, 1898.

'Physics and Sociology', *Contemporary Review*, Vol. LXVIII, Pt. I, 1895; Vol. 69, Pts. II and III, 1896.

Malthus, T. H., *Essay on Population*, First Essay 1798, reprint with notes by James Bonar, Macmillan, 1966.

Marchant, James (ed.), *Alfred Russel Wallace — Letters and Reminiscences* (2 vols), Cassell, 1916.

Marshall, Alfred, 'Distribution and Exchange', *Economic Journal*, Vol. VIII, 1898.

   Principles of Economics, Macmillan, 1890–1907.

   'The Old Generation of Economists and the New', *Quarterly Journal of Economics*, Vol. XI, Jan. 1897.

   The Economics of Industry (1879), Macmillan 1881.

Marx/Engels, *Selected Correspondence*, Progress Publishers, Moscow, 1955.

Massart, J. and Vandervelde, E., *Parisitism, Organic and Social*, Swan Sonnenschein, 1895.

Maudsley, H., *The Physiology and Pathology of the Mind*, 1867.

MacIver, R. M., 'What is Social Psychology?', *Sociological Review*, Vol. VI, 1913.

MacCabe, J., *The Evolution of Mind*, A. & C. Black, 1910.

   Edward Clodd — a Memoir, John Lane, 1932.

MacDonald, T. Ramsay, *Labour and the Empire*, George Allen, 1907.

   Socialism and Society (Aug. 1905), Independent Labour Party Publication, 2nd edn. Oct. 1905.

McDougall, William, *Anthropology and History* (Boyle Lecture No. 22), Oxford University Press, Oxford, 1920.

   'Behaviourism' in J. B. Watson and W. McDougall, *The Battle of Behaviourism*, Kegan Paul, 1928.

   Body and Mind, Methuen, 1911.

   Introduction to Social Psychology, Methuen, 1908.

   The Group Mind, Cambridge University Press, Cambridge, 1920.

   Religion and the Sciences of Life, Methuen, 1934.

   'Modern Materialism — a Reply to Hugh S. Elliott *Bedrock*, Vol. I, 1913.

   'A Practicable Eugenic Suggestion', *Sociological Papers*, 1906.

   'Psychology in the Service of Eugenics', *Eugenics Review*, Vol. V, 1914.

   The Riddle of Life, Methuen, 1938.

McLennan, J. F., *Primitive Marriage*, Edinburgh A. & C. Black, 1865.

   Studies in Ancient History, Macmillan, 1876.

Meldola, Raphael, *Evolution, Darwinian and Spencerian*, (Herbert Spencer Lecture), Clarendon Press, Oxford, 1910.

Merz, J. T., *A History of European Thought in the Nineteenth Century* (4 vols), Blackwoods, Edinburgh, 1896–1914.

Mill, J. S., *Autobiography*, Longmans, 1873.

   A System of Logic, (2 vols), Parker, 1843.

   'The Positive Philosophy of Auguste Comte', *Westminster Review*, (New Series). Vol. XXVIII, April, 1865.

   'Review of Bain's *Senses and Intellect*', *Edinburgh Review*, Vol. CX, 1859.

Mivart, St George, *Contemporary Evolution*, Macmillan, 1876.

   On the Genesis of Species, Macmillan, 1871.

   Review of Bateson's *Materials for the Study of Variation* (1894), *Edinburgh Review*, Vol. CLXXXII, 1895.

Moore, G. E., *Principa Ethica* (1903). Cambridge University Press, Cambridge, 1971.

Morgan, C. Lloyd, 'Darwinism and Psychology' in *Darwinism and Modern Science,* (ed. A. C. Seward, 1909.

Emergent Evolution (The Gifford Lectures, 1922), Williams & Norgate, 1923.

Eugenics and Environment, Bale, 1919.

Instinct and Experience, Methuen, 1912.

An Introduction to Comparative Psychology (The Contemporary Science Series), Walter Scott, 1889.

The Springs of Conduct, Kegan Paul, 1885.

Morgan, Forrest (ed.), *The Life and Works of Walter Bagehot* (5 vols), Travellers Insurance Co. Hartford, Conn., 1889.

Morgan, Lewis, Henry, *Ancient Society,* H. Holt & Co., New York, 1877.

Systems of Consanguinity and Affinity of the Human Family, Smithsonian Institute, Columbia, 1870.

Morgan, T. H., *Evolution and Adaption,* Macmillan, 1903.

Myers, Charles S., 'The Future of Anthropometry', *Journal of the Anthropological Institute,* Vol. XXXIII, 1903, pp. 36–40.

'Instinct and Intelligence – A Reply', *British Journal of Psychology,* Vol. III, 1910.

Pearson, Charles Henry, *National Life and Character. A Forecast,* Macmillan, 1893.

Pearson, Karl, An Address to the Anthropological Section of the British Association, 1920.

'Charles Darwin. An Appreciation'. Delivered to teachers of the L.C.C. *(Questions of the Day and Fray,* No. 12), 1923.

'Darwinism, Medical Progress and Eugenics', Cavendish Lecture, 1912.

National Life from the Standpoint of Science (lecture given 1900), A. & C. Black, 1901.

The Groundwork of Eugenics, Dulau, 1912.

'On the Inheritance of the Mental and Moral Characters in Man and its Comparison with the Inheritance of Physical Characters', Huxley Lecture, 1903. *Journal of the Anthropological Institute,* Vol. XXXIII, 1903.

'On the Relationship of Intelligence to Size and Shape of Head and to other Physical and Mental Characters', *Biometrika,* Vol. V, 1906–7.

'On the Fundamental Conceptions of Biology', *Biometrika,* Vol. I, 1901–2.

'Socialism and Natural Selection', *Fortnightly Review* (New Series), Vol. LVI, July 1894.

The Chances of Death and other Studies, (2 vols), Arnold, 1897.

The Ethic of Free Thought, E. W. Allen, 1883.

The Grammar of Science, (Contemporary Science Series), 1889; A. & C. Black, 1900.

'The Law of Ancestral Heredity', *Biometrika,* Vol. II, 1902–3.

The Life, Letters and Labours of Francis Galton (3 vols), Cambridge University Press, Cambridge, 1914, 1924, 1930.

Pigou, A. C., *The Economics of Welfare,* Macmillan, 1920.

Pitt-Rivers, A. L-F., *The Evolution of Culture and other Essays, Collected and*

Published by Henry Balfour, Clarendon Press, Oxford, 1906.

Plekhanov, G., *In Defence of Materialism* (1895), Lawrence & Wishart, 1947.

Pollock, Frederick, 'Evolution and Ethics', *Mind,* Vol. I, 1876.

Poulton, Edward Bagnell, *Charles Darwin and the Origin of the Species,* Longmans, 1909.

Essays on Evolution 1889–1907, Clarendon Press, Oxford, 1908

'Darwin and Bergson on the Interpretation of Evolution', *Bedrock,* Vol. I, 1912.

Charles Darwin and the Theory of Natural Selection, Cassell, 1896.

Prichard, James Cowles, *The Natural History of Man,* 1843.

Researches into the Physical History of Man, (2 vols) 1813; 2nd edn., J. & A. Arch, 1826.

Quetelet, M. A., *A Treatise on Man* (trans. 1842 by Robert Knox from the French edition of 1835, *Sur l'homme et le development de ses facultes*), Robert & William Chambers, Edinburgh, 1842.

Radl, E., *The History of Biological Theories* (Leipzig, 1905–9), Oxford University Press, Oxford, 1930.

Rentoul, R. R., *Race Culture or Race Suicide?,* Walter Scott, 1906.

Ripley, W., *The Races of Europe,* Kegan Paul, 1900.

Ritchie, D. G., *Darwin and Hegel,* Swan Sonnenschein, 1893.

Darwinism and Politics, Swan Sonnenschein, 1889.

The Principles of State Interference (1891), 4th edn., Swan Sonnenchein, 1902.

'Social Evolution', *International Journal of Ethics',* Vol. VI, 1896.

Review of Bengamin Kidd's *Social Evolution, International Journal of Ethics,* Vol. V, 1894.

'Professor Green's Political Philosophy', *Contemporary Review,* Vol. LI, 1887.

Rivers, W. H. R., *Psychology and Politics,* Kegan Paul, 1923.

Kinship and Social Organisation, Kegan Paul, 1914.

Instinct and the Unconscious, Cambridge University Press, Cambridge, 1920.

The Todas, Macmillan, 1906.

Robertson, J. M., *Buckle and His Critics,* Swan & Sonnenschein, 1895.

Robertson Smith, W., Review of the *History of Human Marriage, Nature,* Vol. LXIV, 23 July, 1891.

Romanes Ethel, (ed.), *The Life and Letters of George John Romanes,* Written and edited by his wife, Longmans, 1896.

Romanes, G. J., *Charles Darwin,* (Nature Series), Macmillan, 1882.

Darwin and After Darwin (3 vols), Longmans, 1892.

Mental Evolution in Animals, Kegan Paul, 1883.

Mental Evolution in Man, Kegan Paul, 1888.

A Review of Aveling's *The Student's Darwin, Nature,* Vol. XXIV, 1881.

'The Darwinian Theory of Instinct', *Proceedings of the Royal Institution,* Vol. XI, 1884.

Ross, E. A., 'Recent Tendencies in Sociology', *Quarterly Journal of Economics,* Vols XVI and XVII, 1902–3.

Russell, E. S., *Form and Function,* John Murray, 1916.

Saleeby, C. W., 'The First Decade of Modern Eugenics', *Sociological Review*, Vol. VII, 1914.

Individualism and Collectivism (Constitutional Issues Series No. 1), Williams & Norgate, 1906.

Parenthood and Race Culture, Cassell, 1909.

The Progress of Eugenics, Cassell, 1914.

Seward, A. C. (ed.), *Darwin and Modern Science*, Cambridge University Press, Cambridge, 1909.

(ed.) *Science and the Nation*, Cambridge University Press, Cambridge, 1917.

Shaw, George Bernard, *Back to Methuselah* (1921), Rev. ed. Oxford University Press, Oxford, 1945.

Sidgwick, Henry, *Miscellaneous Essays and Addresses*, Macmillan, 1904.

'The Relation of Ethics to Sociology', *International Journal of Ethics*, Vol. X, 1899.

The Principles of Political Economy, Macmillan, 1883.

Spencer, Herbert, 'A Theory of Population Deduced from the General Law of Animal Fertility', *Westminster Review* (New Series), Vol. I, 1852.

Autobiography, Williams & Norgate, 1904.

Factors of Organic Evolution, Williams & Norgate, 1887.

First Principles, Williams & Norgate, 1862.

'The Inadequacy of Natural Selection', *Contemporary Review*, Vol. LXIII, 1893.

'Mental Evolution, *Contemporary Review*, Vol. XVII, June, 1871.

Principles of Biology, (2 vols), Williams & Norgate, 1864–67.

Principles of Ethics, Williams & Norgate, 1892.

The Principles of Psychology, (1855) Williams & Norgate, 1864–67.

'Progress, its Law and Cause', *Westminster Review*, (New Series), Vol. XI, 1857.

Reasons for Dissenting from the Philosophy of A. Comte, 1864.

Social Statics, (1851); Williams & Norgate, 1868.

'The Comparative Psychology of Man', *Mind*, Vol. I, No. 1, 1876.

'The Development Hypothesis', *The Leader*, 1857.

'The Social Organism', *Westminster Review* (New Series), Vol. XVII, 1860.

The Man versus the State, (1884), Penguin, 1969.

Spiller, G. 'Darwinism and Sociology', *Sociological Review*, Vol. VII, 1914.

Stephen, Sir James FitzJames, *Liberty, Equality, Fraternity*, Smith, Elder, 1873.

Stephen, Leslie, 'An Attempted Philosophy of History', *Fortnightly Review*, (New Series), Vol. XXVII 1880.

'Belief and Conduct', *Nineteenth Century*, Vol. XXIV, 1888.

'Darwinism and Divinity' in *Essays on Freethinking and Plainspeaking*, Longmans Green & Co., 1873.

'Heredity — an Address to Ethical Societies', in *Social Rights and Duties*, Vol. II, Swan Sonnenschein, 1896.

The Science of Ethics, Smith, Elder, 1882.

'Ethics and the Struggle for Existence', *Contemporary Review*, Vol. LXIV, 1893.

Stout, G., *The Groundwork of Psychology*, Hinds & Noble, New York, 1903.

Sully, James, 'The Relation of the Evolution Hypothesis to Human Psychology' in *Sensation and Intuition*, 1874.

My Life and Friends, Fisher Unwin, 1918.

Sutherland, A. *The Origin and Growth of the Moral Instinct*, (2 vols), Longmans, 1898.

Teggart, F. J., *The Processes of History*, Yale University Press, New Haven, 1918.

Thomson, J. A., *Herbert Spencer*, Green, 1906.

'The Sociological Appeal to Biology', *Sociological Papers*, Vol. III, 1907.

Review of J. Haycraft Berry *Darwinism and Race Progress*, *International Journal of Ethics*, Vol. V, 1895.

Trotter, W., *Instincts of the Herd in Peace and War*, Unwin, 1916.

'The Herd Instinct and its Bearing on the Psychology of Civilised Man', *Sociological Review*, Vol. I, 1908 and Vol. II, 1909.

Tylor, E. B., *Anthropology*, Macmillan, 1881.

'On a Method of Investigating the Development of Institutions: Applied to the Laws of Marriage and Descent', *Journal of the Anthropological Institute*, Vol. XVIII, 1889.

Primitive Culture (2 vols), John Murray, 1871.

Researches into the Early History of Mankind, John Murray, 1865.

Vogt, Karl, *Lectures on Man*, ed. J. Hunt, Publication of the Anthropological Society of London, 1863.

Wake, C. S., *Chapters on Man*, Trubner, 1868.

Wallace, A. R., *A Narrative of Travels on the Amazon*, Reeve, 1853.

Darwinism, Macmillan, 1889.

Contributions to the Theory of Natural Selection, Macmillan, 1870.

My Life — a Record of Events and Opinions, Chapman & Hall, 1905.

Studies Scientific and Social, (2 vols), Macmillan, 1900.

The Malay Archipelago, (2 vols), Macmillan, 1869.

'The Origin of Human Races Deduced from the Theory of Natural Selection', *Anthropological Review*, Vol. II, 1864.

Review of Tylor's *Anthropology*, *Nature*, Vol. XXIV, 1881.

Review of Galton's *Hereditary Genius*, *Nature*, Vol. I, March 1870.

Review of Kidd's *Social Evolution*, *Nature*, Vol. XLIX, 1894.

Social Environment and Moral Progress, Cassell, 1913.

Wallas, Graham, *Human Nature in Politics*, Constable, 1908.

'Instinct'— a Symposium', *British Journal of Psychology*, Vol. III, 1909–10.

Men and Ideas, Allen & Unwin, 1940.

Physical and Social Science (Huxley Memorial Lecture), Macmillan, 1930.

The Great Society, Macmillan, 1914.

Ward, James, *Naturalism and Agnosticism* (Gifford Lecture), A. & C. Black, 1899.

Psychological Principles, The Cambridge Psychological Library, 1915.

Weismann, A. *Studies in the Theory of Descent*, (1875 Leipzig) trans. R. Meldola, (3 parts), Simpson, Law, 1880–2.

Wells, H. E., *The New Machiavelli*, John Lane, 1911.

Westermarck, E. A., 'Sociology as a University Study', 1908, University of London.

'Letters from E. B. Tylor and A. R. Wallace to Edward Westermarck', ed. Rob K. Wikman, *Acta Academiae Aboensis Humaniora XIII*, Vol. XILI, No. 7, 1940.

Memories of my Life, Allen & Unwin, 1929.

The History of Human Marriage (1891) (3 vols).

Whetham, W. E. D. & C. D. 'The Influence of Race on History, in *Problems in Eugenics* Published and edited by the London Eugenics Education Society, 1912.

Wilson, Daniel, *Prehistoric Man*, Cambridge, 1862.

Sources After 1920

Abrams, Philip, *The Origins of British Sociology*, University of Chicago Press, Chicago, 1968.

Addison, P., *The Road to 1945*, (Cape, 1975), Quartet, 1977.

Alland, A., 'Cultural Evolution — the Darwinian Model', *Social Biology*, Vol. XIX, No. 3, Sept. 1972.

Allen, Garland, *The Life Sciences in the Twentieth Century*, Wiley, New York, 1975.

Annan, Noel G., 'The Curious Strength of Positivism in English Thought', Hobhouse Memorial Lecture, 1959.

Leslie Stephen — his Thought and Character in Relation to his Time, McGibben & Kee, 1951.

'The Intellectual Aristocracy in the Nineteenth Century' in *Studies in Social History* ed. J. H. Plumb, Longmans, 1955.

Ardrey, R., *African Genesis* (1961), Fontana/Collins, 8th Imp., 1972.

Arendt, H., *The Origins of Totalitarianism*, Meridan Books, Cleveland, 1951.

Auerbach, *The Science of Genetics* (1962), Hutchinson, 1969.

Avineri, S., 'From Hoax to Dogma', *Encounter*, Vol. XXVIII, March 1967.

Ayala, F. J., 'Teleological Explanations in Evolutionary Biology', *Philosophy of Science*, Vol. XXXVII, March 1970.

Ball, T., 'Marx and Darwin. A Reconsideration', *Political Theory*, Vol. VII, No. 4, 1979.

Bannister, R. C., 'The Survival of the Fittest — History or Histrionics?', *Journal of the History of Ideas*, Vol. XXXI, July/Sept. 1970.

'Sumner's Social Darwinism', *History of Political Economy*, Vol. V, No. 1, 1973.

Banton, M. (ed.), *Darwinism and the Study of Society*, Tavistock, 1961.

Barker, A. D., 'An Approach to the Theory of Natural Selection', *Philosophy*, Vol. XLIV, 1969.

Barker, R., *Political Ideas in Britain*, Methuen, 1978.

Barkow, J. K., 'Culture and Sociobiology' *American Anthropologist*, Vol. LXXX, 1978.

Barnes, H. E. (ed.), *The History and Prospects of the Social Sciences*, Knopf, New York, 1925.

Barnett, S. A. (ed.), *A Century of Darwin*, Heinemann, 1958.

Barnicot, N. A., 'From Darwin to Mendel' in *Man, Race and Darwin*, (Papers read before the Joint Conference of the Royal Anthropological Institute of Great Britain and Ireland and the Institute of Race Relations), 1960.

Barzun, J., *Darwin, Marx and Wagner*, Secker and Warburg; revised edn., Garden City, New York, Doubleday & Co., Anchor Books, 1958 (1942).

Bateson, Beatrice, *William Bateson, A Memoir*, Cambridge University Press, Cambridge, 1928.

Bendix, R., *Work and Authority in Industry*, Harper & Row, New York, 1956.

Bernard, L. L., 'The Misuse of Instinct in the Social Sciences', *Psychological Review*, Vol. XXVIII, 1921.

Bibby, Cyril, *T. H. Huxley — Scientist, Humanist and Educator*, Watts, 1959.
'Huxley and the Origin', *Victorian Studies*, Vol. III, 1959.

Biddiss, M. (ed.), *Gobineau — Selected Political Writings*, Cape, 1971.
Images of Race, (The Victorian Library), Leicester University Press, 1979.
Gobineau — Father of Racist Ideology, Weidenfeld & Nicolson, 1970.

Birdsell, Joseph, B., 'The Problem of the Evolution of the Human Race', *Social Biology*, Vol. XVIII, No. 2, June 1972.

Blacker, C. P., *Eugenics — Galton and After*, Duckworth, 1952.
Birth Control and the State, Kegan Paul, 1926.
'Eugenics in Germany', *Eugenics Review*, Vol. XXV, no. 3, 1933.

Boardman, P., *The Worlds of Patrick Geddes*, Routledge & Kegan Paul, 1978.

Bock, Kenneth E., 'The Acceptance of Histories' in W. J. Cahnman, and A. Boskoff, *Sociology and History*, Collier-Macmillan, Free Press of Glencoe, New York, 1964.
'Darwinism and Social Theory', *Philosophy of Science*, Vol. XXII, 1955.

Bodelson, C. A., Studies in Mid Victorian Imperialism, Gyldendalske Boghandel, København, 1924.

Boden, M., *Purposive Explanation in Psychology*, Harvard University Press, Cambridge, Mass., 1972; Harvester Press, Hassocks, 1978.
Piaget, Harvester Press, Brighton, 1979.

Bogardus, E. S., *A History of Social Thought*, University of Southern California Press, Los Angeles, 1922.

Bolt, Christine, *Victorian Attitudes to Race*, Routledge & Kegan Paul, 1971.

Boring, E. G., *A History of Experimental Psychology*, Century Co., New York, 1929.
'Measurement in Psychology', *Isis*, Vol. LII, 1961.

Bowle, John, *Politics and Opinion in the Nineteenth Century*, Cape, 1954.

Bowler, P. J., 'The Changing Meaning of Evolution, *Journal of the History of Ideas*, Vol. XXXVI, 1975.
'Malthus, Darwin and the Concept of Struggle', *Journal of the History of Ideas*, Vol. XXXVII, 1976.

Brew, J. O., *One Hundred Years of Anthropology*, Harvard University Press, Cambridge, Mass., 1968.

Buchan, A. F., *The Spare Chancellor — the Life of Walter Bagehot*, Chatto & Windus, 1959.

Burrow, J. W., 'Evolution and Anthropology in the 1860s — the Anthropological Society of London', *Victorian Studies*, Vol. VII, 1963.

Evolution and Society (1966), Cambridge University Press, Cambridge, 1970.

'Uses of Philology' in *Ideas and Institutions of Victorian Britain*, ed. R. Robson, Bell, 1967.

Cahnman, W. T. and Boskoff, A. (ed), *Sociology and History*, Collier-Macmillan, Free Press of Glencoe, New York, 1964.

Canfield, John, 'Teleological Explanation in Biology', *British Journal for the Philosophy of Science*, Vol. XIV, 1964.

Canguilhem, G., *La Formation du concept de réflexe aux viie et xviiie siecles'*, Paris, Presses Universitaires de France, 1955.

Etudes d'Histoire de Philosophie des Sciences, Paris, VRIN, 1968.

La Connaissance de la Vie, Paris VRIN, 1952.

Clarke, Peter, *Liberals and Social Democrats*, Cambridge University Press, Cambridge, 1978.

Cole, G. D. H. *Samuel Butler*, Longmans, Green, 1952.

Collini, S., *Liberalism and Sociology*, Cambridge University Press, Cambridge, 1979.

Connolly, Kevin, 'The Concept of Evolution — a Comment on Papers by Mr Manser and Professor Flew', *Philosophy*, Vol. XLI, 1966.

Cowles, T., 'Malthus, Darwin and Bagehot— a Study in the Transference of a Concept', *Isis*, Vol. XXVI, 1937.

Cox, O. C., *Caste, Class and Race*, Doubleday, New York, 1948.

Crook, D. P., 'Was Benjamin Kidd a Racist?', *Ethnic and Racial Studies*, Vol. II, No. 2, 1979.

Cullen, M. J., *The Statistical Movement in Early Victorian Britain*, Harvester Press, Hassocks, 1975.

Curtis, L. P., 'Anglo Saxons and Celts — a Study of Anti-Irish Prejudice', Bridgeport Conn, Conference on British Studies, 1968.

Apes and Angels — the Irishman in Victorian Caricature, David & Charles, Newton Abbot, 1971.

Daniel, Glyn, *A Hundred Years of Anthropology*, Duckworth, 1950.

Davis, Kingsley, *Human Society* Macmillan 1949; 20th edn. Macmillan, New York, 1965.

Dawkins, R., *The Selfish Gene*, Oxford University Press, Oxford, 1976.

De Beer, Sir Gavin, *Charles Darwin*, Nelson, 1963.

Delattre, F., *Samuel Butler et le Bergsonisme*, Paris, 1936.

Dobzhansky, T., *Evolution, Genetics and Man*, (1955) Wiley, New York, 1971.

Driver, C. H., 'Walter Bagehot and the Social Psychologists' in *Social and Political Ideas of some Representative Thinkers of the Victorian Age*, ed. F. J. C. Hearnshaw, Harrap, 1933.

'The Development of a Psychological Approach to Politics in English Speculation', Appendix to *Social and Political Ideas of some Representative Thinkers of the Victorian Age'*, (ed.) F. J. C. Hearnshaw, Harrap, 1933.

Dunn, L. C., *A Short History of Genetics*, McGraw-Hill, New York, 1965.

Durant, J. R., 'Scientific Naturalism and Social Reform in the Thought of

Alfred Russel Wallace', *British Journal of the History of Science*, Vol. XII, No. 40, March 1979.

Durkheim, E., *Rules of Sociological Method*, (Paris, 1895), Glencoe Press, 1950.

Eiseley, Loren, *Darwin's Century — Evolution and the Men who Discovered it*, Gollancz, 1959.

Ellegard, Alvar, *Darwin and the General Reader, 1859–72*, Universitets arsskrift, Goteborg, 1958.

'The Darwinian Theory and Nineteenth Century Philosophies of Science', *Journal of the History of Ideas*, Vol. XVIII. 1957.

Evans, Pritchard, et al, *Essays in Honour of C. G. Seligmann*, Kegan Paul, 1934.

Evans, R. I., *Konrad Lorenz: The Man and His Ideas*, Harcourt Brace Jovanovich, New York, 1975.

Farrall, Lyndsey A., 'The Origins and Growth of the English Eugenic Movement 1865–1925', Indiana University Doctoral Dissertation, 1970, University Microfilms 70–14964.

Feuer, Lewis S., 'The Friendship of Edwin Ray Lankester and Karl Marx', *Journal of the History of Ideas*, Vol. XL, no. 4, 1979.

Feyeraband, P., *Against Method*, NLB, 1975.

Science in a Free Society, NLB, 1978.

Firth, R. W. (ed.), *Man and Culture*, Routledge, 1957.

Fisher, R. A., *The Genetical Theory of Natural Selection*, Clarendon Press, Oxford, 1930.

Fletcher, R., *The Making of Sociology*, Vol. 1, Michael Joseph, 1971.

John Stuart Mill, Michael Joseph, 1971.

Flew, A., *Evolutionary Ethics*, (1967), Macmillan, 1970.

'The Structure of Malthus's Theory of Population', *Australian Journal of Philosophy*, Vol. XXXV, 1957.

Flugel, J. C., *A Hundred Years of Psychology 1833–1933*, Duckworth, 1933.

Forest, D. W., *Francis Galton: The Life and Work of a Victorian Genius*, Elek, 1974.

Fortes, M., 'Social Anthropology in Cambridge since 1900', Inaugural Lecture at Cambridge University, Cambridge, 1955.

Fox, R. (ed.), *Biosocial Anthropology*, Malaby Press, 1975.

Freeden, M., *The New Liberalism*, Clarendon Press, Oxford, 1978.

Gasman, D. V., *The Scientific Origins of National Socialism — Social Darwinism Ernst Haeckel and the German Monist League*, MacDonald; London, New York, American Elsevier, 1971.

George, Wilma, *Biologist, Philosopher. A Study of the Life and Writings of Alfred Russell Wallace*, Abelard-Schuman, 1964.

Gerrantana, V., 'Marx and Darwin', *New Left Review*, No. 82, Nov./Dec. 1973.

Ghiselin, M., *The Triumph of the Darwinian Method*, University of California Press, Berkeley, 1969.

Gillespie, Neal C., *Charles Darwin and the Problem of Creation*, University of Chicago Press, Chicago, 1979.

Gillispie, C. C., *Genesis and Geology*, Harvard University Press, Cambridge, Mass., 1951.

The Edge of Objectivity, Princeton University Press, 1960.

'The Sociology of Science', Congress Internationale de l'Histoire des Sciences, Paris, 1968.

Ginsberg, M., *Essays in Sociology and Social Philosophy*, (1947) (2 vols), Heinemann, 1961.

Studies in Sociology, Methuen, 1932.

Glass, B., Temkin, O. and Strauss, W. L., *Forerunners of Darwin 1745–1859* (1959), Johns Hopkins University Press, Baltimore, 1968.

Glick, T. F., (ed.), *The Comparative Reception of Darwinism*, University of Texas Press, Austin, 1974.

Goldman, Irving, 'Evolution and Anthropology', *Victorian Studies*, Vol. III, 1959.

Gossett, T. F., *Race — the History of an Idea in America*, Southern Methodist University Press, Dallas, 1963.

Greene, J. C., *The Death of Adam and its Impact on Western Thought*, Iowa State University Press, Ames, Iowa, 1959.

Griffin, C. M., 'A Critical Examination of L. T. Hobhouse's Social and Political Theories', PhD Thesis, University of London 1972.

Grinder, Robert E., *A History of Genetic Psychology*, Wiley, New York, 1967.

Gruber, H. E., 'Darwin and Das Kapital', *Isis*, Vol. LII, 1961.

Darwin on Man, Wildwood House, 1974.

Gruber, H. E. and Varèche, J. J. (ed.), *The Essential Piaget*, Routledge & Kegan Paul, 1977.

Gurvich, C. and Moore, W., *La Sociologie au XXe siecle*, Paris, 1947.

deGuistino, D., *Conquest of Mind*, Croom Helm, 1975.

Hadfield, J. A. (ed.), *Psychology and Modern Problems*, University of London Press, 1935.

Haldane, J. B. S., 'Communication', *Science and Society*, No. 5, 1941.

'Karl Pearson', speech delivered at University College London on the occasion of the Karl Pearson Centenary Celebrations, 1957.

Heredity and Politics, Allen & Unwin, 1938.

Hall, R., *Marie Stopes*, Deutsch, 1977.

Haller, J. S., *Outcasts from Evolution: Scientific Attitudes of Racial Inferiority. 1859–1900*, University of Illinois Press, Urbana, Illinois, 1971.

Haller, M., *Eugenics*, Rutgers University Press, 1963.

Halliday, R. J., 'Social Darwinism: A Definition', *Victorian Studies*, Vol. XIV, 1971.

'The Sociological Movement, the Sociological Society and the Genesis of Academic Sociology in Britain', *Sociological Review* (New Series), Vol. XVI, 1968.

Harré, R. (ed.), *The Rationality of Scientific Revolutions*, Clarendon Press, Oxford, 1974.

Harris, Marvin, *The Rise of Anthropological Theory*, Routledge & Kegan Paul, 1968.

Harvie, C., *The Lights of Liberalism; University Liberal's and the Challenge of Democracy*, Allen Lane, 1976.

Hawkins, D., (ed.), *D. H. Lawrence. Stories, Essays and Poems* (1939) Dent, 1969.

Hawthorn, G., *Enlightenment and Despair*, Cambridge University Press, Cambridge, 1976.

Hays, H. R., *From Ape to Angel — an Informal History of Social Anthropology*, Knopf, New York, 1958.

Hearnshaw, F. J. C. (ed.), *Social and Political Ideas of some Representative Thinkers of the Victorian Age*, Harrap, 1933.

Hearnshaw, L. S., *A Short History of British Psychology, 1840–1940*, Methuen, 1964.

   Cyril Burt, Psychologist, Hodder & Stoughton, 1979.

Hearnshaw, L. S. and Watson, R. I., 'Differential Psychology' in *Congress Internationale de l'historiens des Sciences*, Paris, 1968.

Herskovits, M. J., *Cultural Dynamics* (1947), Knopf, New York, 1967.

Hilgard, E., 'Psychology After Darwin' in *Evolution After Darwin*, ed. Tax Sol, 1960.

Himmelfarb, G., *Darwin and the Darwinian Revolution*, Chatto & Windus, 1959.

   'Malthus', *Encounter*, Vol. v, 1955.

Hinde, R. A., *Biological Bases of Human Social Behaviour*, McGraw-Hill, New York, 1974

Hobson, J. A. and Ginsberg, M. (eds.), *L. T. Hobhouse — His Life and Work*, Allen & Unwin, 1931.

Hodgen, M. T., *The Doctrine of Survivals*, Allenson, 1936.

Hogben, Lancelot, 'The Race Concept', and 'Man, Race and Darwin' (Papers read at the Joint Conference of the Royal Anthropological Institute of Great Britain and Ireland and the Institute of Race Relations), 1960.

Hofstadter, R., *Social Darwinism in American Thought* (1944), revised edn., New York, Braziller, 1965.

Horkheimer, M., *Eclipse of Reason*, Oxford University Press, Oxford, 1947.

Houghton, W. E., *The Victorian Frame of Mind* (1957), Yale University Press, New Haven, 1974.

Howard, D. T., 'The Influence of the Evolution Doctrine on Psychology', *Psychological Review*, Vol. XXXIV, 1927.

Hull, D. L., *Darwin and his Critics — the Reception of Darwin's Theory of Evolution by the Scientific Community*, Harvard University Press, Cambridge, Mass., 1973.

   Philosophy of Biological Science, Prentice-Hall, New Jersey, 1974.

Hulse, Fred, S., 'Technological Advance and Major Racial Stocks', *Human Biology*, Vol. XXVII, 1955.

   'Adaption, Selection and Plasticity in Ongoing Human Evolution', *Human Biology*, Vol. XXXII, 1960.

   'Natural Selection and Differential Population Growth of the Human Races', *Social Biology*, Vol. XIX, No. 2, 1972.

Huxley, J. S., *Essays of a Biologist*, Chatto & Windus, 1923.

Huxley, J. S. and Haddon, A. C., *We Europeans*, Cape, 1935.

Irvine, W., *Apes, Angels and Victorians*, Weidenfeld & Nicolson, 1955.

Jarvie, I. C., *The Revolution in Anthropology*, Routledge & Kegan Paul, 1964.

Jay, Martin, *The Dialectical Imagination*, Heinemann, 1973.

Johnson, H. M., *Sociology — a Systematic Introduction* (1961), 6th Imp., Routledge & Kegan Paul, 1968.

Jones, Greta, J., 'The Social History of Darwin's *Descent of Man*', *Economy and Society*, Vol. VII, No. I, Feb. 1978.

'Lysenko, British Scientists and the Cold War', *Economy and Society*, Vol. VIII, No. 1, Feb. 1979.

'Il Darwinismo Nella Cultural Inglese e Americana Studi Recenti', *Studi Storici*, Vol. XIX, 1978.

Keith, Sir Arthur, *An Autobiography*, Watts, 1950.

Kent, John, *From Darwin to Blatchford — the Role of Darwinism in Christian Apologetics 1875–1910*, Dr Williams Trust, 1966.

Keynes, J. M., *Alfred Marshall, 1842–1924 — a Memoir*, 1924.

Kiernan, V. G., *The Lords of Human Kind* (1969), Penguin, 1972.

Kitchen, P., *A Most Unsettling Person: An Introduction to the Ideas and Life of Patrick Geddes*, Gollancz, 1975.

Kolakowski, L., *Positivist Philosophy* (Doubleday, USA, 1968), Penguin, 1972.

Krautz, L. and Allen, D., 'The Rise and Fall of McDougall and Instinct', *Journal of Behavioural Science*, Vol. III, No. 4, Oct. 1967.

Kroeber, A. L., *Anthropology, Biology and Race*, Harbinger Books, New York, 1963.

Kuhn, T. S., *The Structure of Scientific Revolutions*, (1962), University of Chicago Press, Chicago, 1970.

Lakatos, I. and Musgrave, L., *Criticism and the Growth of Knowledge*, Cambridge University Press, Cambridge, 1970.

Lazarfield, P., 'Notes on the History of Quantification in *Sociology*', *Isis*, Vol. LIII, 1961.

Lecourt, D., *Marxism and Epistemology* (1969), NLB, 1975.

Lerner, I. M., and Libby, W. J., *Heredity, Evolution and Society*, W. H. Freeman, San Francisco, 1976.

Levin, S., 'Malthus and the Idea of Progress', *Journal of the History of Ideas*, Vol. XXVII, Jan./March 1966. ·

Lichtenberger, J. P., *Development of Social Theory*, Century, New York, 1923.

Limoges, C., *La Selection Naturelle*, Presses Universitaires de France, Paris, 1970.

Lippincott, B. E., *Victorian Critics of Democracy*, University of Minnesota Press, Minneapolis, 1938.

Loewenberg, B. J., *Darwin, Wallace and the Theory of Natural Selection*, Arlington Books, Cambridge, Mass., 1959.

Lorenz, K., *On Aggression*, Methuen, 1966.
  Civilised Man's Eight Deadly Sins, Methuen, 1973.

Lovejoy, A. O., *The Great Chain of Being*, Harvard University Press, Cambridge, Mass., 1936.

Lowie, R. H., *The History of Ethnological Thought*, Harrap, 1938.

MacBeth, N., *Darwin Retried* (1971), Garnstone Press, 1974.

MacKenzie, W. J. M., *Biological Ideas in Politics*, Penguin, 1978.

MacRae, D. G., *Ideology and Society*, Heinemann, 1961.
  'Darwinism and the Social Sciences', in *A Century of Darwin*, ed. S. A. Barnett, 1958.
  Introduction to *Man Versus the State — a Collection of Articles by Herbert*

Spencer, Penguin, 1969.

McConnaughey, G., 'Social Darwinism', *Osiris,* Vol. IX, 1950.

McKinney, H. L., *Wallace and Natural Selection,* Yale University Press, New Haven, 1972.

Malinowski, B., *A Scientific Theory of Culture and other Essays* (1944), Oxford University Press, Oxford, 1969.

Manier, E., *The Young Darwin and His Cultural Circle,* Reidel, Dordecht and Boston, 1978.

Manser, A. R., 'The Concept of Evolution', *Philosophy,* Vol. XL, 1965.

Marett, R. R., 'James George Frazer', Proceedings of the British Academy, 1941.

Tylor, Chapman & Hall, 1936.

Maude, A., *The Life of Marie Stopes,* Peter Davies, 1933.

Meek, R. (ed.), *Marx and Engels on Malthus,* Lawrence & Wishart, 1953.

Merton, R. K., *Science, Technology and Society in Seventeenth Century England,* Osiris History of Science Monographs, Bruges, 1938.

Midgley, M., *Beast and Man,* Harvester Press, Sussex, 1979.

Millhauser, Milton, *Just Before Darwin,* Wesleyan University Press, Middleton, Conn., 1959.

Montagu, Ashley, 'Communication', *Science and Society,* Vol. VI, 1942.

'Karl Pearson and the Historical Method in Anthropology', *Isis.* Vol. XXXIV, 1943.

The Nature of Human Aggression, Oxford University Press, New York, 1976.

Moore, James R., *The Post-Darwinian Controversies,* Cambridge University Press, Cambridge, 1979.

Murchison, C. (ed.), *A History of Psychology in Autobiography,* Vol. II, Clark University Press, Worcester, Mass., 1930.

Murphy, G., *An Historical Introduction to Modern Psychology,* Kegan Paul, 1929.

Murray, R. H., *Studies in the English Social and Political Thinkers of the Nineteenth Century* (2 vols), Heffer, Cambridge, 1929.

Nisbet, A., *Konrad Lorenz,* Dent, 1976.

Nisbet, R., *Social Change and History,* Oxford University Press, New York, 1969.

Nordenskiöld, N. E., *The History of Biology,* Kegan Paul, 1929.

Norton, B. J., 'The Evolutionary Theory of the Biometricans', M. Phil (unpublished) London University, 1971.

Parsons, T., *The Structure of Social Action,* (1937) (2 vols), The Free Press of Glencoe, Collier-Macmillan, 1968.

Passmore, J., *The Perfectibility of Man,* Duckworth, 1970.

Pearson, E. S., *Karl Pearson,* Cambridge University Press, Cambridge, 1938.

Peel, J. D. Y., *Herbert Spencer. The Making of a Sociologist,* Heinemann, 1971.

Peirce, C. S., *Collected Papers of Charles Sanders Peirce,* ed. C. Hartshorne & P. Weiss, Cambridge, Mass. Harvard University Press, 1931–58.

Penniman, T. K., *A Hundred Years of Anthropology,* Duckworth, 1935.

Phillips, D. C., 'Organicism in late Nineteenth and early Twentieth Century Thought', *Journal of the History of Ideas,* Vol. XXXI, 1970.

Bibliography 215

Piaget, J., *Psychology and Epistemology* (1970), Allen Lane, Penguin, 1972.
Biology and Knowledge, Edinburgh University Press, Edinburgh, 1971.
The Principles of Genetic Epistemology, (Presses Universitaires de France, 1970) Routledge & Kegan Paul, 1972.
Polyani, M., 'The Republic of Science', *Minerva*, Vol. I, No. 1, 1962.
Popper, K., 'Evolution and the Tree of Knowledge' in *Objective Knowledge*, Oxford University Press, Oxford, 1972.
Conjectures and Refutations, Routledge & Kegan Paul, 1963.
Unended Quest (1974), Fontana/Collins, 1977.
The Logic of Scientific Discovery, Hutchinson, 1959.
Provine, W. B., *The Origins of Theoretical Population Genetics*, University of Chicago Press, Chicago, 1971.
Punnett, R. C., 'Early Days of Genetics', *Heredity*, Vol. IV, pt. 1, April 1950.
Quiggin, A. H., *Haddon–Headhunter*, Cambridge, University Press, 1942.
Quillian, W. F., *The Moral Theory of Evolutionary Naturalism*, Yale University Press, New Haven, 1945.
Randell, J. H., 'The Changing Impact of Darwinism on Philosophy', *Journal of the History of Ideas*, Vol. XXII, Oct./Dec. 1961.
Richards, Robert J., 'Lloyd Morgan's Theory of Instinct from Darwinism to Neo-Darwinism', *Journal of the History of the Behavioural Sciences*, Vol. XIII, 1977.
'Influence of Sensationalist Tradition on Early Theories of the Evolution of Behaviour', *Journal of the History of Ideas*, Vol. XL, No. 1, 1979.
Richter, Melvin, *The Politics of Conscience — T. H. Green and His Age*, Weidenfeld & Nicolson, 1964.
Roach, J., 'Liberalism and the Victorian Intelligentsia', *Cambridge Historical Journal*, Vols. XXII, Oct./Dec. 1961.
Robinson, A. L., *William McDougall* (Memorial Lecture), Dukes University Press, South Carolina, 1938.
Rogers, James A., 'Darwinism and Social Darwinism', *Journal of the History of Ideas*, Vol. XXXIII, May/June 1972.
Rosenberg, C. E., *No Other Goods* (1961), Johns Hopkins, University Press, Baltimore, 1976.
Royle, E., *Victorian Infidels*, Manchester University Press, Manchester, 1974.
Rumney, J., *Herbert Spencer's Sociology* Williams & Norgate, 1934.
'British Sociology' in Twentieth Century Sociology, ed. G. Gurvich and W. Moore, 1945. New York, Philosophical Library.
Ruse, M., *The Philosophy of Biology*, Hutchinson University Library, 1973.
The Darwinian Revolution, University of Chicago Press, Chicago, 1979.
Sociobiology: Sense or Nonsense?, Reidel, Dordrecht/Boston, 1979.
Russett, C. E., *The Concept of Equilibrium in American Social Thought*, Yale University press, New Haven, 1966.
Darwin in America, San Francisco, W. H. Freeman, 1976.
Sahlins, M., *The Use and Abuse of Biology* (1976), Tavistock, 1977.
Sandow, A., 'Chance and Necessity — an Essay on the Natural Philosophy of Modern Biology', *Science and Society*, Vol. XXVII, 1972.
Schiller, F. C. S., 'Eugenics and the Press' *Eugenics Review*, Vol. XX, 1928–9.
Searle, G. R., *The Quest for National Efficiency*, Blackwell, Oxford, 1971.

Eugenics and Politics in Britain 1900–14, Noordhof International, Leyden, 1976.

'Eugenics and Politics in Britain in the 1930s', *Annals of Science*, Vol. XXVI, 1979.

Seligmann, C. G., 'Anthropology and Psychology'. *Journal of the Anthropological Institute*, Vol. LIV, 1924.

Semmel, B., *Imperialism and Social Reform*, Allen & Unwin, 1960.

The Governor Eyre Controversy, McGibbon & Kee, 1962.

Shaw, G. B., *Back to Methuselah*, (1921) Revised ed., Oxford University Press, 1945.

Skinner, B. F., *Beyond Freedom and Dignity*, Cape, 1972.

About Behaviourism, Cape, 1974.

Reflections on Behaviourism and Society, Prentice-Hall, New Jersey, 1978.

Skinner, G. D., 'The Limits of Historical Explanation', *Philosophy*, Vol. XLI, 1966.

Soffer, R., Ethics and Society in England: The Revolution in the Social Sciences, University of California Press, Berkeley, 1978.

Somit, A. (ed.), *Biology and Politics*, Mouton, The Hague/Paris, 1976.

Stanton, William, *The Leopard's Spots — Scientific Attitudes towards Race in America 1815–59*, University of Chicago Press, Chicago, 1960.

Stark, A., 'Would Darwin's Theory have been Different if he had known of Mendel's Law?' (Unpublished M.Phil), London University, 1972.

Staude, J. R., *Max Scheler 1874–1928*, Collier-Macmillan, New York, 1967.

Stern, B. J., 'Communication', *Science and Society*, Vol. V, 1941.

Stocking, G. W., *Race, Culture and Evolution*, Collier-Macmillan, New York, 1969.

'Lamarckianism in American Social Thought 1890–1915', *Journal of the History of Ideas*, Vol. XXIII, June 1962.

Stokes, E., 'The Political Ideas of English Imperialism', Inaugural Lecture given in the University College of Rhodesia and Nyasaland, 1960.

Swinbourne, R. G., 'Galton's Law — Foundation and Development', *Annals of Science*, Vol. XXI, 1965.

Taton, Rene, (ed.), *A General History of the Sciences*, Thames and Hudson, 1966.

Tax, Sol (ed.), *Evolution After Darwin*, University of Chicago Press, Chicago, 1960.

Vorzimmer, P. J., 'Charles Darwin and Blending Inheritance', *Isis*, Vol. LIV, 1963.

'Darwin's Ecology and its Influence on his Theory', *Isis*, Vol. LVI, 1965, *Charles Darwin — the Years of Controversy* (1970), University of London Press, London, 1972.

Waddington, C. H., *The Strategy of the Genes*, Allen & Unwin, 1957.

The Ethical Animal, Allen & Unwin, 1960.

Webb, Beatrice, *My Apprenticeship* (1926), reprinted by Penguin, 1971.

Wersky, P., *The Visible College*, Allen Lane, 1978.

Wichler, Gerhard, *Charles Darwin*, Pergamon Press, Oxford, 1961.

Wiener, M., *Between Two Worlds — The Political Thought of Graham Wallas*, Clarendon Press, Oxford, 1971.

Wiener, P., *Evolution and the Founders of Pragmatism,* (Harvard, 1949), Peter Smith, Gloucester, Mass., 1969.

Wilkie, J. S., 'Some Reasons for the Rediscovery of Mendel's Work in the First Years of the Present Century', *British Journal of the History of Science,* Vol. I, 1962–3.

Willey, B., *Darwin and Butler,* Chatto & Windus, 1960.

Williams, Raymond, 'Social Darwinism' in *The Limits of Human Nature,* ed. Jonathan Benthall, Allen Lane, 1973.

Wilson, E. O., *Sociobiology. The New Synthesis,* Harvard University Press, Cambridge, Mass., 1975.

On Human Nature, Harvard University Press, Cambridge, 1978.

'Biology and the Social Sciences'. *Daedalus,* Vol. II 1977.

'Some Central problems of Sociobiology', *Social Sciences Information,* Vol. XIV (6) 1975.

Wiltshire, D., *The Social and Political Thought of Herbert Spencer,* Oxford University Press, Oxford, 1978.

Young, R. M., *Mind, Brain and Adaption in the Nineteenth Century,* Clarendon Press, Oxford, 1970.

'Malthus and the Evolutionists', *Past and Present,* Vol. XLIII, No. 43, 1968.

'Darwin's Metaphor— does Nature Select?' *The Monist,* Vol. LV, 1971.

'The Impact of Darwin on Conventional Thought' in *The Victorian Crisis of Faith,* A. Symundson, SPCK (Society for the Promotion of Christian Knowledge), 1970.

Zirkle, Conway, 'Malthus, Benjamin Franklin and the US Census', *Isis,* Vol. XLVIII, 1957.

Evolution, Marxian Biology and the Social Scene, University of Pennsylvania Press, Philadelphia, 1959.

'Gregor Mendel and his Precursors', *Isis,* Vol. XLII, 1951.

NOTES

Preface

1 W. J. M. Mackenzie, *Biological Ideas in Politics,* 1978, p. 35.
2 R. Hofstadter, *Social Darwinism in American Thought,* 1944. Criticisms of his interpretation of the influence of social Darwinism can be found in R. C. Bannister, 'Sumner's Social Darwinism' *History of Political Economy,* Vol. V, No. 1, 1973, and C. E. Russett, *Darwin in America,* 1976, pp. 93–6.
3 Quoted in Gruber, *Darwin on Man,* 1974, p. 317.

Chapter I

1 A. Comte, *Positive Philosophy* (1853), Vol. II, p. 112.
2 Benjamin Kidd, interview in the *Daily Chronicle,* 20 June, 1894.
3 W. Bagehot, 'Malthus' in *The Life and Works of Walter Bagehot,* ed. Forrest Morgan, Vol. V, p. 396.
4 D. G. Ritchie, 'Social Evolution', *International Journal of Ethics,* Vol. VI, No. 2, Jan. 1896, p. 165.
5 A. Marshall, *The Economics of Industry* (1879), 1881, p. 9.
6 J. S. Mill, *A System of Logic,* 1843, Vol. II, p. 499.
7 A. Marshall, 'The Old Generation of Economists and the New', *Quarterly Journal of Economics'* Vol. XI, Jan. 1897, p. 118.
8 J. Merz, *A History of European Thought in the Nineteenth Century,* 1912, Vol. III, p. 607.
9 W. H. Mallock, 'Physics and Sociology', *Contemporary Review,* Vol. LXIX, 1896, p. 60.
10 Grant Allen, *Charles Darwin,* 1885, p. 186.
11 Reverend William Kirby, 'On the History, Habits and Instincts of Animals, in *The Bridgewater Treatises* (1835), Vol. I, p. 142.
12 J. Crawfurd, 'On the Theory of the Origin of Species by Natural Selection', *Transactions of the Ethnological Society of London,* 1869, Vol. VII, p. 29.
13 H. Spencer, 'A Theory of Population deduced from the General Law of Animal Fertility', *Westminster Review,* (New Series), Vol. I, No. 2, 1852, pp. 468–501.
14 W. R. Greg, 'On the Failure of Natural Selection in the Case of Man', *Fraser's Magazine,* Vol. LXXVIII, Sept. 1868, p. 356.
15 K. Pearson, 'Darwinism and Medical Progress', the Cavendish Lecture, 1912, p. 11 (my italics).
16 T. H. Huxley, 'The Struggle for Existence: A Programme', *Nineteenth Century,* Vol. XXIII, 1888, p. 169.
17 E. Nordenskiöld, *History of Biology,* 1929, p. 470.

Chapter II

1 Compiled, 1837–40. They include the following: 1837a–1838a, Transmutation Notebook B; 183b, Transmutation Notebook C; 1838c Transmutation Notebook D; 1838d–1839a, Transmutation Notebook E; 1838e, Metaphysical Notebook M; 1838f–1839b, Metaphysical Notebook N; 1837b–1840a, 'Old and Useless Notes about the Moral Sense and some Metaphysical Points written about the Year 1837 and Earlier.' 1838g, 'On Macculloch, Attributes of the Deity', The Darwin Collection, Anderson Room, University Library, Cambridge University, Cambridge, England.

2 Sir A. Grant, 'Philosophy and Mr Darwin', *Contemporary Review*, Vol. XVII, May, 1871, p. 179.

3 Darwin, N. Notebook in *Darwin on Men*, ed. H. E. Gruber, 1974, p.342.

4 G. J. Romanes, *Mental Evolution in Man*, 1888, p. 7.

5 Darwin, M. Notebook in H. E. Gruber ed., 1974, p. 295.

6 J. S. Mill, Review of Bain's *Senses and Intellect*, *Edinburgh Review*, Vol. CX, 1859, p. 224.

7 H. Maudsley, *Physiology and Pathology of Mind*, 1867, pp. 110–11.

8 C. Darwin, *Descent of Man*, 1871, Vol. I, p. 35.

9 H. Spencer, *Principles of Psychology*, 1855, pp. 491–2.

10 Spencer to Youmans, 5 June 1871, in D. Duncan (ed.), 1908, p. 149 (the article referred to appeared in the *Contemporary Review*, June 1871).

11 Quoted in C. Limoges, *La Selection Naturelle*, 1970, p. 82 fn.

12 H. Spencer, *Principles of Psychology*, 1855, p. 492.

13 J. Lubbock, 'On the Origin and Early Condition of Man', Geography and Ethnology Section, *British Association*, Dundee, 1867, Vol. XXXVII, pp. 118–25.

14 C. Darwin, *Descent of Man*, 1871, Vol. I, p. 181, *ibid.* p. 182; *ibid.* p. 181.

15 E. B. Tylor, Preface to 2nd ed. of *Primitive Culture* (1871), 1873, pp. vii–viii.

16 A. Marshall, *Principles of Economics* (1890), 8th ed., 1947, Appendix C., p. 772.

17 L. Stephen, 'An Attempted Philosophy of History', *Fortnightly Review*, Vol. XVII, 1880, p. 678.

18 C. Darwin, *The Descent of Man*, 1871, Vol. I, pp. 71–2.

19 W. K. Clifford, *Lectures and Essays*, 1879, p. 106.

20 C. Darwin, *The Descent of Man*, 1871, Vol. I, p. 167.

21 Hyndman to Wallace, Letter of 14 March, 1912, in *Letters and Reminiscences of A. R. Wallace*, ed. J. R. Marchant, Vol. II, 1916, p. 164.

22 Jno. Rhind, 'The Survival of the Fittest', *The Social Democrat*, 15 December, 1910, Vol. XIV, No. 12, p. 545–546.

23 Wallace, Letter to Mr A. Wiltshire, 14 Sept. 1913, in Marchant (ed.), *op. cit.*, Vol. II, p. 165.

24 A. R. Wallace, *Malay Archipelago*, Vol. I, p. 141; *Travels on the Amazon*, p. 172; *Malay Archipelago*, *op. cit.*, p. 144.

25 A. R. Wallace, *My Life*, p. 87.

26 A. R. Wallace, *Travels on the Amazon*, p. 85.

27 A. R. Wallace, 'The Origin of Human Races and the Antiquity of Man deduced from the Theory of Natural Selection', *Anthropological Review*, Vol. II, March 1, 1864, p. clxii.

28 A. R. Wallace, *My Life*, 1905, Vol. II, p. 17.

29 A. R. Wallace, *Contributions to the Theory of Natural Selection*, p. 359; *ibid.* p. 319; *ibid.* p. 318.

30 Wallace to Darwin, 2 July, 1866, in Marchant (ed.), *op. cit.*, p. 170.

31 A. R. Wallace, Review of Tylor's *Anthropology, Nature*, Vol. XXIV, 1881, p. 242.

32 A. R. Wallace in British Association Meeting, reported in *The Times*, 25 Aug. 1869.

Chapter III

1 See B. Semmel, *The Governor Eyre Controversy*, 1962 and C. Harvie, *The Lights of Liberalism; University Liberals and the Challenge of Democracy*, 1976.

2 See W. R. Greg, 'On the Failure of Natural Selection in the case of Man', *Fraser's Magazine*, Vol. LXXVIII, 1868, and F. Galton, 'Hereditary Talent and Character', Pts 1 and 2, *Macmillan's Magazine*, Vol. XII, Pts. 1 and 2, 1865.

3 Charles Darwin, Letter to J. Hooker, 25 Jan. 1862 in *Life and Letters of Charles Darwin*, ed. Francis Darwin, 1887, Vol. II, p. 385.

4 F. Galton, *Hereditary Genius*, 1869, p. 362.

5 D. G. Ritchie, 'Professor Green's Political Philosophy', *Contemporary Review*, Vol. LI, June 1887, p. 846.

6 J. S. Mill, *Autobiography*, 1873, p. 225; *ibid.* p. 225, *ibid.* pp. 273–4.

7 F. Pollock, 'Evolution and Ethics', *Mind*, Vol. I, No. 3, 1876, p. 337; *ibid.* p. 344; F. Pollock in the introduction to W. K. Clifford, *Lectures and Essays*, ed. Leslie Stephen and F. Pollock, 1879, p. 33.

8 W. K. Clifford, 'Right and Wrong', p. 169, reprinted in *Lectures and Essays*, 1879, Vol. II, p. 169, (from the *Fortnightly Review*, Vol. II, Dec. 1875); W. K. Clifford, 'On the Scientific Basis of Morals', *ibid.* Vol. II, p. 106 (reprinted from *Contemporary Review*, Vol. XXVI, 1875).

9 C. Darwin, *The Descent of Man*, 1871, Vol. I, pp. 161–2.

10 Leslie Stephen, Letter to B. J. Norton, 16 March, 1877, in '*The Life and Letters of Leslie Stephen*, ed. F. W. Maitland, 1906, p. 300.

11 W. Bagehot, 'Letters on the French Coup d'Etat', 20 Jan., 1852, in *The Works of Walter Bagehot*, ed. Forrest Morgan, 1889, Vol. II, p. 379.

12 L. Stephen, *The Science of Ethics*, 1882, p. 124.

13 K. Popper, 'Evolution and the Tree of Knowledge' in *Objective Knowledge*, 1972, p. 261.

14 M. Polyani, 'The Republic of Science', *Minerva*, Vol. I, No. 1, 1962.

15 H. Sidgwick, 'The Relation of Ethics to Sociology', *International Journal of Ethics*, Vol. X, 1899. For Green, See M. Richter *T. H. Green and His Age*, 1964.

16 A. Sutherland, *The Origin and Growth of the Moral Instinct* (2 Vols), 1898, Vol. I, p. vii.

17 See S. Alexander, *Moral Order and Progress,* 1889; D. G. Ritchie, *Darwinism and Politics,* 1889, *Darwin and Hegel,* 1893.

18 F. W. Maitland (ed.), *The Life and Letters of Leslie Stephen,* 1906, p. 326.

19 See L. T. Hobhouse, 'The Diversions of a Psychologist' in *The Pilot,* Vol. V, No. 97, 4 Jan. 1902; L. T. Hobhouse, *Development and Purpose* (1913), Preface to 1927 ed., p. xx.

20 J. S. Mill, *Autobiography,* 1873, p. 273.

21 L. T. Hobhouse's Syllabus on Sociology, London School of Economics Calendar 1907–9, p. 147; *ibid.* 1909–10, p. 165.

22 L. T. Hobhouse in *The Pilot, op. cit.* p. 13.

23 A. Marshall, *Principles of Economics,* 1890, Vol. I, p. 303.

24 L. Stephen, 'Ethics and the Struggle for Existence', *Contemporary Review,* Vol. LXIV, Aug. 1893, p. 168; *ibid.* pp. 168–9.

25 B. Bosanquet, 'Socialism and Natural Selection', lecture given before the London Ethical Society. Published in *Aspects of the Social Problem,* 1895, p. 299.

26 L. Stephen, 'Heredity — an Address to Ethical Societies' in *Social Rights and Duties,* Vol. II, 1896, p. 53.

27 N. G. Annan, 'The Intellectual Aristocracy in the Nineteenth Century' in *Studies in Social History,* ed. J. H. Plumb, 1955.

28 J. S. Mill, *A System of Logic,* 1843, Vol. II, pp. 606–7; *ibid.* p. 607.

Chapter IV

1 Beatrice Webb, *My Apprenticeship* (1926). Reprinted by Penguin, 1971, pp. 186–7.

2 Charles Darwin, Letter to C. Lyell. 4 Jan. 1860, in *The Life and Letters of Charles Darwin,* ed. F. Darwin, 1887, Vol. II, p. 262.

3 L. T. Hobhouse, *Development and Purpose,* 1927, 2nd edn, (1st edn, 1913), p. xvii.

4 See H. Spencer, *The Man versus the State,* 1884. (Essays reprinted from the *Contemporary Review* Feb.–July 1884); H. Spencer, Letter to Count Goblet D'Alviella, 7 Jan., 1895, in *The Life and Letters of Herbert Spencer,* ed. D. Duncan, 1908, p. 336.

5 P. Geddes, *An Analysis of the Principles of Economics,* Pt. 1, 1885. Read before the Royal Society of Edinburgh, 7 July 1884, p. 36; *ibid.* pp. 5–6; *ibid.* p. 28.

6 W. H. R. Rivers, 'The Morbid in Sociology' in *Psychology and Politics,* 1923, pp. 63–4.

7 D. G. Ritchie, 'Mr Spencer's Individualism and his Conception of Society', (First published in the *Contemporary Review,* 1886), in *Principles of State Intervention,* 1902, 4th edn, p. 50; *ibid.* p. 15.

8 T. H. Huxley, 'The Struggle for Existence — a Programme', *Nineteenth Century,* Vol. XXIII, Feb. 1888, and 'Evolution and Ethics', the Romanes Lecture 1893.

9 D. G. Ritchie, *Darwin and Hegel,* 1893, p. 63. D. G. Ritchie 'Review of Benjamin Kidd's *Social Evolution*', *International Journal of Ethics,* Vol. V, No. 1, 1894–5, pp. 110–11.

10 L. Stephen, 'Ethics and the Struggle for Existence', *Contemporary Review*, Vol. LXIV, 1893, p. 165.

11 L. T. Hobhouse, *The Labour Movement*, 1893, p. xi; *ibid.* p. 91; *ibid.* pp. 91–2.

12 C. W. Saleeby, *Individualism and Collectivism* (Constitutional Issues Series), 1906, No. 1, p. 129.

13 See E. Aveling, *The Religious Views of Charles Darwin*, 1883, *The Student's Darwin*, 1881, and *The Gospel of Evolution*, 1884.

14 Karl Marx, Letter to Kugelmann, 27 June, 1870, in K. Marx and F. Engels, *Selected Correspondence* (1955), 1965 p. 240.

15 G. Plekhanov, *In Defence of Materialism*, 1947, p. 146.

16 See E. Royle, *Victorian Infidels*, 1974, Ch. 3.

17 F. Engels, Letter to Lavrov, 12–17 Nov. 1875, *op. cit.*, p. 301.

18 F. Engels, *Anti-Duhring* (1878), 1969, 5th edn. (1st edn, 1947), p. 84; *ibid.* p. 89; *ibid.* p. 116; *ibid.* p. 118; *ibid.* p. 117.

19 P. LaFargue, 'A Few Words with Mr Spencer', *Today*, No. 5 Jan./June, 1884, p. 417.

20 *Justice*, 13 June, 1885, p. 1; W. Willis-Harris, 'The Survival of the Fittest', in *Justice* 28 April, 1888, p. 2.

21 P. Geddes, 'On the Application of Biology to Economics', *Meeting of the British Association*, Aberdeen, 1885, Vol. LV, p. 1167.

22 A. P. Hazell, 'Co-operation versus Socialism', *Justice*, 17 March, 1888.

23 J. Massart and E. Vandervelde, *Parasitism, Organic and Social*, 1895, p. 25; *ibid.* p.v. (see Geddes's preface).

24 E. Ferri, *Socialism and Positive Science*, 1905, (trans. from the French edition of 1896 by Edith C. Harvey, Originally published 1894).

25 J. Addison, 'The Survival of the Fittest in Politics' Pt. II of *The Social Democrat*, 1910, Vol. XIV, No. 7, pp. 300–1.

26 J. Keir Hardie, *From Serfdom to Socialism*, 1907, p. 92; *ibid.* p. 94; *ibid.* pp. 87–8.

27 J. R. MacDonald, *Socialism and Society*, 1905, 2nd edn, p. 13.

28 Robert Rives La Monte, 'Science and Revolution', *The Social Democrat*, Vol. XIII, No. 3, 15 March, 1909, p. 111; J. Addison, *ibid.* p. 299; Robert Rives La Monte, *ibid.* p. 105.

29 J. R. MacDonald, *op. cit.*, p. 112 fn.

30 Quoted from *The Anarchist*, in *Justice*, Dec. 1886, p. 4.

Chapter V

1 See Garland Allen, *The Life Sciences in the Twentieth Century*, 1975.

2 See P. J. Vorzimmer, *Charles Darwin — the Years of Controversy*, (1970) 1972.

3 C. Darwin, *Descent of Man*, 1871, Vol. I, p. 103.

4 H. Spencer, Preface to *The Factors of Organic Evolution*, 1887, p. iv (articles which, first appeared in *Nineteenth Century*, Vol. XIX, 1886).

5 W. Bagehot, 'Letters on the French Coup d'Etat', 29 Jan, 1852 in *The Works of Walter Bagehot*, ed. Forrest Morgan, 1889, Vol II, p. 401; *ibid.* p. 395.

6 R. H. Hutton, 'Memoir of Walter Bagehot', *ibid.* Vol. I, p. liii.

7 W. Bagehot, *Physics and Politics,* 1872, p. 8 (quoted from H. Maudsley *The Physiology and Pathology of Mind,* 1867); *ibid.* p. 8.

8 C. Darwin, *Descent of Man,* 1871, Vol. I, p. 104.

9 C. Darwin, Letter to Mrs Talbot, 19 July 1881, from *More Letters of Charles Darwin,* ed. Francis Darwin, 1903, Vol. II, p. 55.

10 W. B. Carpenter, 'The Hereditary Transmission of Acquired Psychical Habits', Part I, *Contemporary Review,* Vol. XXI, Jan. 1873, p. 306.

11 F. Pollock, 'Evolution and Ethics', *Mind,* Vol. I, 1876, p. 337.

12 A. Weismann, *Studies in the Theory of Descent,* 1882 (trans. Ralph Meldola).

13 L. T. Hobhouse, 'The Value and Limitations of Eugenics' in *Social Evolution and Political Theory,* Columbia University Press, New York, 1911, Ch. 3, p. 64.

14 Wallace to Meldola, 10 June 1893, in *Alfred Russel Wallace — Letters and Reminiscences,* ed. J. Marchant, 1916, Vol. II, p. 56.

15 H. Spencer, 'The Inadequacy of Natural Selection', *Contemporary Review,* Vol. LXIII, March 1893, (and Weismann/Spencer controversy, *Contemporary Review,* Vol. LXIII and LXIV); H. Spencer, *The Factors of Organic Evolution,* (1887) p. 9; *ibid.* p. 29.

16 Kropotkin, 'The Direct Action of Environment on Plants', *Nineteenth Century,* Vol. LXVIII, July 1910, p. 61.

17 G. J. Romanes, 'The Darwinian Theory of Instinct', *Proceedings of the Royal Institution,* Vol. XI, 1884, p. 136.

18 J. M. Baldwin, Letter to Wallace, 9 Nov. 1902, in *Between Two Wars,* J. M. Baldwin, 1926, Vol. XI, p. 248; Wallace to Baldwin, 15 Aug. 1902, *ibid.* p. 246.

19 St G. Mivart, *On the Genesis of Species,* 1871, p. 230; *ibid.* pp. 230–1.

20 G. B. Shaw, Preface to *Back to Methuselah* (1921), rev. edn, 1945, p. xvii (see also Samuel Butler, *Evolution, Old and New,* 1879, *Life and Habit,* 1877; *Luck or Cunning,* 1887).

21 See W. Bateson, *Mendel's Principles of Heredity,* 1902. For the integration of Darwinism and Mendelian genetics, see, among others, R. A. Fisher, *The Genetical Theory of Natural Selection,* 1930.

22 Letter from Wallace to Poulton, 8 Sept. 1894, in Marchant (ed.) *op. cit.,* Vol. II, pp. 60–1. Letter of Wallace to Sir J. Hooker, 10 Nov. 1905, *ibid.* p. 82.

23 E. B. Poulton, 'Mutation, Mendelism and Natural Selection' in *Essays on Evolution,* 1908, pp. xvii–xviii.

24 Quoted in Floris Delattre, *Samuel Butler et le Bergsonisme,* Paris 1936, p. 396.

25 H. Bergson, *Creative Evolution* (1907), trans. 1911, 2nd edn, Nov. 1911, p. 81; *ibid.* p. 81.

26 Letter from Graham Wallas to George Bernard Shaw, 1921, *Shaw Papers* (British Museum), ADD M.S.S. 50553 undated.

27 W. Bateson, *Evolution and Education,* 1922 (Address written 1915), pp. 421–2 in B. Bateson *William Bateson, A Memoir,* (1928).

28 P. Kropotkin, 'The Direct Action of Environment on Plants', *op. cit.* p. 77 fn.

29 C. W. Saleeby, *The Progress of Eugenics*, 1914, pp. 148–9.

30 Hugh S. Elliott, Review of J. Loeb's, *The Mechanistic Conception of Life*, *Bedrock*, Vol. II, No. 1, April 1913, pp. 122–4. E. B. Poulton, 'Darwin and Bergson on the Interpretation of Evolution', *ibid*. Vol. I, No. 1, April 1912, p. 49.

31 See Hugh S. Elliott, *Modern Science and the Illusions of Professor Bergson*, 1912.

32 Quoted from McDougall, *Body and Mind*, 1911, p. 257, in Hugh S. Elliott, 'Modern Vitalism', *Bedrock*, Vol. I, No. 1, Oct. 1912, p. 319; *ibid*. p. 320.

33 Saleeby, *op. cit.*, p. 148.

34 W. McDougall, 'Was Darwin Wrong?' (*The Forum*, 1928), reprinted in *Religion and the Sciences of Life*, 1934, pp. 172–85.

35 G. B. Shaw, 'The Lysenko Muddle', *Labour Monthly*, Jan, 1949, p. 18; *ibid*. p. 19.

Chapter VI

1 Address to the National Association for the Promotion of Social Science, quoted as 'Social Science', *Blackwood's Magazine*, Vol. XC, Oct. 1861, p. 464.

2 F. Galton, quoted in *The Life, Letters and Labours of Francis Galton* (3 vols), ed. Karl Rearson, Vol. II, 1924, p. 86.

3 F. Galton 'Hereditary Improvement', *Fraser's Magazine*, (New Series), Vol. III, 1873, pp. 117–18.

4 C. Darwin, *Descent of Man*, 1871, Vol. I, p. 177–8; *ibid*. p. 178.

5 C. Darwin, Letter to Galton, 23 Dec. 1870 in *More Letters of Charles Darwin*, ed. F. Darwin, 1903, Vol. II, p. 41.

6 F. Galton, *Heredity Genius*, 1869, p. 46; *ibid*. p. 9.

7 W. R. Greg 'On the Failure of Natural Selection in the Case of Man', *Fraser's Magazine*, Vol. LXXVIII, , 1868, p. 361; *ibid. p. 360; ibid. p. 360.*

8 *Report of the Committee on Physical Deterioration*, 1904, Vol. XXXII, (cd, 2175 Parliamentary Papers.) para. 2267, p. 102.

9 J. S. Mill, *Autobiography*, 1873, p. 231.

10 M. A. Quetelet, *A Treatise on Man*, 1842, p. 97; *ibid*. p. 88.

11 See D. deGiustino, *Conquest of Mind*, 1975.

12 See John Beddoe, *The Races of Britain — A Contribution to the Anthropology of Western Europe*, 1885; Also the *Journal of the Anthropological Institute*, 1907; Recommendations of the Committee on Physical Deterioration, 1904; A. C. Haddon, *Reports of the Cambridge Expedition to the Torres Straits* Vol. II, 'Physiology and Psychology', 1901; A. H. Quiggin, *Haddon – Headhunter*, Cambridge University Press, Cambridge, 1942. Also see Galton's contribution to anthropometry and anthropology in the *Journal of the Anthropological Institute* in the 1890s.

13 A. C. Haddon *The Study of Man*, 1898, p. 63.

14 See Mill's attack in *A System of Logic*, 1843, Vol. II, p. 499 and A. Bain in *On the Study of Character including an Estimate of Phrenology*, 1861.

15 F. Galton, *Heredity Genius*, 1869, p. 2.

16 Charles S. Myers, 'The Future of Anthropometry', *Journal of the Anthropological Institute*, Vol. XXXII, 1903, p. 40.

17 K. Pearson, 'On the Relationship of Intelligence to Size and Shape of Head and to other Physical and Mental Characters', *Biometrika*, Vol. V, 1906–7, p. 133; *ibid.* p. 136.

18 K. Pearson, 'On the Inheritance of the Mental and Moral Characters in Man and its Comparison with the Inheritance of Physical Characters' (Huxley Lecture, 1903, *Journal of the Anthropological Institute*, 1903, Vol. XXXIII,, p. 193.

19 F. Galton, 'Eugenics, Its Definition, Scope and Aims', *Sociological Papers*, 16 May, Vol. I, 1905, p. 50.

20 L. Stephen, 'Heredity — an Address to Ethical Societies' in *Social Rights and Duties*, 1896, Vol. II, p. 21.

21 B. Kidd, *The Science of Power*, 1918, p. 77.

22 *Report of the Committee on Physical Deterioration*, 1904, *op. cit*, para. 282, p. 55.

23 Quoted in C. W. Saleeby, *Parenthood and Race Culture*, 1909, p. 158 fn.

24 See Havelock Ellis, 'The Problem of Race Regeneration' in *New Tracts for the Times*, 1911.

25 A. R. Wallace, 'Evolution and Character', *Fortnightly Review*, Vol. LXXXIII, Jan. 1908. Quoted in C. W. Saleeby, *The Eugenic Prospect*, 1921, p. 194.

26 Gissing based a number of his feminist novels on similar themes of sexual choice proscribed by considerations of income and social status, e.g. *The Odd Women*. His own life was also dogged by similar problems.

27 C. W. Saleeby, *Parenthood and Race Culture*, 1909, p. 165.

28 S. Webb, 'Eugenics and the Poor Law — Minority Report'. Delivered to the Eugenics Society, 1909 in *Papers on the Report of the Poor Law Commission*, Eugenics Education Society, 1911; *ibid.* p. 240.

29 H. J. Laski, 'The Scope of Eugenics', *Westminster Review*, Vol. LXXIV, 1910, pp. 25–34.

30 A. R. Wallace, Letter to C. G. Stuart-Menteith, 5 June, 1901 in Alfred Russel Wallace – *Letters and Reminiscences*, ed. J. Marchant, 1916, Vol. II, p. 160.

31 L. T. Hobhouse, 'The Value and Limitations of Eugenics' in *Social Evolution and Political Theory*, 1911, p. 46–7.

32 K. Pearson, *The Groundwork of Eugenics*, 1912, p. 33.

33 Investigation into Pauper Family Histories, Eugenics Education Society. *Papers on the Report of the Poor Law Commission*, 1911, sect. 3, pp. 187–8.

34 A. Marshall, *The Principles of Economics* (1890–1907), 8th edn, 1947, p. 201.

35 A. C. Pigou, *The Economics of Welfare*, 1920, p. 91; *ibid.* p. 98.

36 See the *Journal of the Anthropological Institute*, 1907 and *Report of the Committee on Physical Detioration*, 1904, *op. cit.* p. 424.

37 *Report of the Commission on the Feeble Minded*, 1908, Parliamentary Papers, (d. 4202) Vol. XXXIX, Pt. 6, p. 181, para. 545.

38 John Gray, *Report of the Committee on Physical Deterioration*, 1904, para. 3267, p. 141, Vol. XXXII, *ibid.* Cunningham, para 2271, p. 102.

39 C. W. Saleeby, 'The First Decade of Modern Eugenics', *Sociological Review,* Vol. VII, 1914; pp. 133–4.

40 K. Pearson. Quoted in E. S. Pearson (1938), pp. 58–9. The reference is to C. B. Davenport's, *Heredity and Eugenics.*

41 See Bateson's comments in the introduction to *Mendel's Principles of Heredity,* 1906. Also see R. C. Punnett, 'Early Days of Genetics', *Heredity,* Vol. IV, Pt. 1, April 1950.

42 W. Bateson, 'Biological Fact and the Structure of Society', Herbert Spencer Lecture, 1912, p. 31.

43 W. Robertson Smith, 'Review of the History of Human Marriage', *'Nature',* Vol. LXIV, 23 July 1891, p. 270.

44 A. R. Wallace, Letter to Westermarck, 20 Jan. 1892, *Acta Academiae Aboensis* (Humaniora XIII, 1940), Vol. XIII, No. 7, p. 21.

Chapter VII

1 B. Kidd, *Social Evolution,* 1894, p. 102.

2 *The Spectator,* 22 Feb. 1902, p. 292.

3 See H. M. Cecil *Pseudo-Philosophy at the End of Nineteenth Century,* 1897. T. H. Huxley, Letter to Edward Clodd, 3 Nov. 1894, in *'The Life and Letters of T. H. Huxley,* ed. Leonard Huxley, 1903, Vol. III, p. 336; L. Stephen, 'Heredity — an Address to Ethical Societies' in *Social Rights and Duties,* 1896, pp. 30–1.

4 D. G. Ritchie, Review of Benjamin Kidd's *Social Evolution, International Journal of Ethics,* Vol. V, No. 1, Oct. 1894, p. 115.

5 J. Hobson, *The Psychology of Jingoism,* 1901, p. 18.

6 G. Wallas, *Human Nature in Politics,* 1908, pp. 105–6; *ibid.* p. 25.

7 See. W. H. R. Rivers, 'Instinct in Relation to Society' in *Psychology and Politics,* 1923; E. A. Westermarck, *Memories of My Life,* 1929, p. 77.

8 C. Darwin, *Origin of Species* (1859), Penguin edition ed. J. W. Burrow, 1968, p. 234 (reprinted 1970); C. Darwin, *Descent of Man,* 1871, Vol. I, p. 37; *ibid.* p. 36.

9 See E. A. Westermarck, *The History of Human Marriage,* (3 vols), 1891.

10 C. Letourneau, *Property its Origin and Development,* 1892, p. 2; *ibid.* p. 5.

11 W. H. R. Rivers, 'Instinct in Relation to Society', in *Psychology and Politics,* 1923, p. 34; W. H. R. Rivers, 'Socialism and Human Nature', *ibid.* p. 86.

12 W. McDougall, *Introduction to Social Psychology,* 1908, p. 181.

13 J. A. Hobson, *op. cit.,* 1901, p. 20.

14 M. Scheler. Quoted in J. R. Staude, *Max Scheler, 1874–1928,* 1967, p. 214.

15 D. H. Lawrence, Letter to Ernest Collinge, 17 Jan. 1913 in *D. H. Lawrence, Poems and Essays,* ed. Desmond Hawkins, (1939), 1969 edn, p. 371.

16 W. McDougall, *The Group Mind,* 1920, p. 100.

17 Sir Arthur Keith, Boyle Lecture 1919, quoted in *Autobiography,* 1950, pp. 398–9.

18 Sociology Examination, Pt. 1, Q. 2, 1910, LSE, set by W. H. R. Rivers

and E. A. Westermarck.

19 See Darwin, in the Appendix to G. J. Romanes, *Mental Evolution in Animals*, 1883; W. Kirby, *Bridgewater Treatise*, 1835, Vol. II, , No. 7, p. 22; see also, G. J. Romanes, *On the Darwinian Theory of Instinct*, Royal Institution, 1884; Lloyd Morgan, see *Introduction to Comparative Psychology*, 1889.

20 C. S. Myers, 'Instinct and Intelligence — a Reply', *British Journal of Psychology*, Vol. III, 1909–10, Oct. 1910, Pt. 3, pp. 267–8.

21 According to McDougall, Freud's work, 'betrays a conception of the nature of the sex instinct which is vague, chaotic and elusive, uncontrolled by consideration of the facts of animal instinct and inconsistent with these facts. In support of this . . . it may suffice to point out that the Freudian conception of the nature and development of sexuality is radically incompatible with the view that the sex impulse is directed towards the opposite sex by the innate organisation of the instinct'. *Introduction to Social Psychology* (1908), Suppl. to Ch. 11, 29th edn, 1948 p. 352.

22 See W. H. R. Rivers, *Instinct and the Unconscious*, 1920.

23 See G. Wallas, 'Instinct and the Unconscious'; the Symposium on Instinct and the Unconscious included W. H. R. Rivers, C. S. Myers, Jung, Wallas, James Drever and W. McDougall; *British Journal of Psychology*, Vol. X, pt. 1, Nov. 1919.

24 See L. Krautz and D. Allen, 'The Rise and Fall of McDougall and Instinct', *Journal of the Behavioural Sciences*, Vol. III, No. 4, Oct., 1967, pp. 326–38.

25 W. McDougall in *British Journal of Psychology*, Vol. III, 1909–10, Oct. 1910, pt. 3, p. 259.

26 L. T. Hobhouse, *The Rational Good*, 1921, pp. 19–20.

27 J. S. Mill, *Autobiography*, 1873, p. 225.

28 L. Stephen, *The Science of Ethics*, 1882, Introduction, p. vi; L. Stephen, 'An Attempted Philosophy of History, *Fortnightly Review*, (New Series), Vol. XXVII, 1880, p. 677; *ibid.* p. 678; *ibid.* p. 678, *ibid.* p. 678.

29 See M. Boden, *Purposive Explanation in Psychology*, 1972.

30 For Lorenz, See Alec Nisbett, *Konrad Lorenz*, 1976, and Lorenz, *On Aggression*, 1966, *Civilised Man's Eight Deadly Sins*, 1974, K. Lorenz in *Konrad Lorenz: The Man and his Ideas*, ed. Richard L. Evans, 1975, p. 128.

31 Mary Midgley, *Beast and Man*, 1979, p. 331.

32 See B. F. Skinner in *The Rationality of Scientific Revolutions*, ed. R. Harré, 1974.

33 J. S. Mill, *Autobiography*, 1873, p. 274.

Chapter VIII

1 See B. Semmel, *The Governor Eyre Controversy*, 1962.

2 F. Darwin, ed., *Life and Letters of Charles Darwin*, 1887, Vol III, p. 53.

3 See J. Crawfurd, 'The Plurality of the Faces of Man', 1867. Address given 13 January at St. Martin's Hall and reprinted in *Proceedings of the Ethnological Society of London*. J. Hunt 'On the Application of the

Principle of Natural Selection in Anthropology', *Anthropological Review*, Vol. IV, 1866.

4 C. S. Wake, *Chapters on Man*, 1868, p. 106.

5 J. S. Mill, *Autobiography*, 1873, p. 274.

6 E. B. Tylor, *Primitive Culture* (1871), 3rd edn, 1891, p. 7.

7 A. C. Haddon, *The Practical Value of Anthropology*, 1921, p. 39.

8 F. Galton, 'Hereditary Talent and Character', *Macmillans Magazine*, Vol. XII, 1865, p.326.

9 See M. Biddiss, Gobineau — Father of Racist Ideology, 1970.

10 A. Marshall, *The Economics of Industry* (1879), 1881, 2nd edn, p. 31; *ibid.* p. 37; *ibid.* pp. 37–8.

11 C. Lloyd Morgan, *The Springs of Conduct*, 1885, p. 241.

12 A. C. Haddon, *The Study of Man*, 1898, p. xxviii; *ibid.* p. xxviii.

13 A.R.Wallace, *Social Environment and Moral Progress*, 1913, p. 34; *ibid.* p.77.

14 W. E. D. and C. D. Whetham, 'The Influence of Race on History', International Conference on Eugenics, 24–30 July 1912, published by the London Eugenics Education Society as *Problems in Eugenics*, 1912, p. 244; *ibid.* p. 246.

15 See R. Rentoul, *Race Culture or Race Suicide?*, 1906.

16 E. A. Westermarck, *Sociology as a University Study*, 1908, p. 31 (Inauguration of the Martin White professorships at the University of London); *ibid.* p. 31.

17 W. H. R. Rivers, 'The Government of Subject Peoples' in *Science and the Nation*, ed. A. C. Seward, 1917, pp. 306–7.

18 F. Galton, 'Ethnological Inquiries on the Innate Character and Intelligence of Different Races'. Quoted in K. Pearson, *The Life, Letters and Labours of Francis Galton*, 1924, Vol. II, pp. 352–3.

19 Sociology Examination, Paper 2, Q.1, 1907, LSE, set by W. H. R. Rivers and E. A. Westermarck; LSE Calendar 1905–6, p. 56; Sociology (Ethnology) Pt. 11, Q.1, 1911, LSE, set by W. H. R. Rivers and C. G. Seligmann; Sociology, Paper 1, Q.3, 1920, LSE, set by E. A. Westermarck, L. T. Hobhouse and A. E. Crawley; Sociology (Internal) Q.7, 1918, LSE, set by E. J. Urwick; Sociology, Pt. 2, Q.3, 1907, LSE, set by W. H. R. Rivers and E. A. Westermarck; Sociology Paper 1, Q.2, 1908, LSE; *ibid.*, Q.8 (my italics).

20 L. Stephen, 'Ethics and the Struggle for Existence', *Contemporary Review*, Vol. LXIV, Aug. 1893, p. 168.

21 B. Kidd, *The Science of Power*, 1918, p. 294.

22 Sociology Examination Pt. 1, Q.4, 1911, LSE, set by L. T. Hobhouse and W. H. R. Rivers.

23 J. R. MacDonald, *Labour and the Empire*, 1907, pp. 98–9.

24 See J. E. Cairnes, 'The Negro Suffrage', *Macmillan's Magazine*, Vol. XII, 1865, pp. 334–43.

25 L. Stephen, *op. cit.* p. 166.

26 T. H. Huxley, 'Emancipation Black and White' in *Lay Sermons*, 1870, pp. 23–4 (from *The Reader*, 20 May 1865).

27 L. T. Hobhouse, *Liberalism*, 1911, p. 43; *ibid.* p. 44.

28 C. Harvey, *The Biology of British Politics*, 1904, p. 115–6.

Chapter IX

1 B. Malinowski, *A Scientific Theory of Culture and Other Essays* (1944), 1969, p. 26.
2 T. Parsons, 'Malinowski and the Theory of Social Systems' in *Man and Culture*, ed. R. Firth, (1957) 1970, pp. 64–5.
3 Malinowski, *op. cit.*, p. 16; *ibid.* pp. 143–4.
4 Kingsley Davis, *Human Society*, 1949, 20th edn, 1965, p. 144.
5 H. M. Johnson, *Sociology — a Systematic Introduction* (1961), 1968, p. 633.
6 M. J. Herskovits, *Cultural Dynamics*, (1947), 1967, pp. 153–4.
7 T. Parsons, 'Evolutionary Universals in Society', *American Sociological Review*, Vol. XXIX, 1964, pp. 339–57; *ibid.* p. 341.
8 See R. Bendix, *Work and Authority in Industry*, 1956.
9 Parsons, *op. cit.*, p. 341.
10 L. Stephen, *The Science of Ethics*, 1882, p. 80.
11 C. P. Blacker, *Birth Control and the State*, 1926, pp. 33–4.
12 See A. Maude, *The Life of Marie Stopes*, 1924; Ruth Hall, *Marie Stopes*, 1977.
13 Editorial on 'Nazi Eugenics', *Eugenics Review*, Vol. XXV, No. 2, July 1933–4, p. 77. C. P. Blacker, 'Eugenics in Germany', *ibid.* Vol. XXV, No. 3, Oct. 1933, p. 157.
14 P. G. Wersky, *The Visible College*, 1978.
15 F. C. S. Schiller, in *Eugenics Review*, Vol. XX, 1928–9, p. 105.
16 See Paul Addison, *The Road to 1945*, (1975), 1977.
17 See Cyril Burt, 'Autobiography' in *A History of Psychology in Autobiography*, ed. C. Murchison, 1952.
18 For C. G. Seligmann see A. C. Haddon, 'Appreciation of C. G. Seligmann' in Evans Pritchard *et al*, *Essays in Honour of C. G. Seligmann*, 1934.
19 C. G. Seligmann in *Psychology and Modern Problems*, ed. J. A. Hadfield, 1935, p. 98.
20 See C. G. Seligmann, 'Anthropology and Psychology. A Study of Some Points of Contact', *Journal of the Anthropological Institute*, Vol. LIV, 1924, p. 13.
21 R. Ardrey, *African Genesis*, (1961) 8th edn. 1972 p. 162.
23 Garland Allen, *The Life Sciences in the Twentieth Century*, 1975.

Chapter X

1 T. H. Huxley (1896), p. 72. Quoted in D. L. Hull, *Darwin and his Critics*, 1973, p. 28.
2 J. S. Mill to Alexander Bain, 11 April 1860, quoted in *John Stuart Mill*, ed. R. Fletcher, 1971, p. 390.
3 E. Nordenskiöld, *History of Biology*, 1929, p. 470.
4 See K. Popper, *The Logic of Scientific Discovery*, 1959. *Conjectures and Refutations*, 1963, *Objective Knowledge* 1972.
5 See Michael Ghiselin, *The Triumph of the Darwinian Method*, 1969.
6 K. Popper *Unended Quest*, (1974), 3rd imp. 1977, p. 168.

7 Georges Canguilhem, 'La Formation du concept de réflexe aux XVIIᵉ et XVIIIᵉ Siecles, 1955.

8 See T. S. Kuhn, *The Structure of Scientific Revolution*, 1962.

9 P. Feyeraband, *Against Method*, 1975, p. 146.

10 M. Ruse, *The Philosophy of Biology*, 1973, p. 218.

11 D. L. Hull, *Darwin and his Critics*, 1973, p. 76.

12 William James, *Principles of Psychology*, (2 vols.) 1890.

13 C. S. Peirce, *Collected Papers of Charles Sanders Peirce*, 1935, Vol. VI.

14 H. Bergson, *Creative Evolution* (1911), 2nd edn, 1911, p. 234. and p. 236.

15 K. Popper, *Objective Knowledge*, 1972, p. 229; *ibid*. p. 278; *ibid*. p. 222.

16 Mivart, *Cosmic Philosophy*, Vol. II, p. 475. Quoted in P. Wiener, *Evolution and the Founders of Pragmatism*, 1949, p. 148.

17 K. Popper, *Objective Knowledge*, 1972, pp. 247–8.

18 E. O. Wilson, *Sociobiology, the New Synthesis*, 1975, p. 4.

19 Jerome K. Barkow, 'Culture and Sociobiology', *American Anthropologist*, LXXX, 1978, p. 13.

20 A. Comte, *Positive Philosophy*, trans. Harriet Martineau, 1853, Vol. II, p. 112.

21 E. O. Wilson, 'Biology and the Social Sciences', *Daedalus*, Vol. II, Autumn 1977, p. 136; E. O. Wilson, *op. cit.* 1975, p. 549.

22 E. O. Wilson, 'Some Central Problems of Sociobiology', *Social Sciences Information*, 14(6), 1975, p. 9.

23 W. Bagehot *Collected Works*, ed. Forrest Morgan, 1889, Vol. IV, p. 433.

24 See Mary Midgley, *Beast and Man*, 1979.

25 A. Comte, *Positive Philsophy*, 1853, Vol. II, p. 115; *ibid*. p. 113; *ibid*. p. 113.

Index

231